# Science

## a children's encyclopedia

# Science
## a children's encyclopedia

**Authors** Chris Woodford, Steve Parker

**Senior Editor** Daniel Mills
**Senior Art Editor** Vicky Short
**Art Editor** Mandy Earey
**Managing Editor** Paula Regan
**Managing Art Editor** Owen Peyton Jones
**Pre-production Producer** Nikoleta Parasaki
**Production Controller** Mary Slater
**Jacket Designer** Laura Brim
**Jackets Editor** Maud Whatley
**Jacket Design Development Manager** Sophia MTT
**Publisher** Sarah Larter
**Art Director** Phil Ormerod
**Associate Publishing Director** Liz Wheeler
**Publishing Director** Jonathan Metcalf

**DK India**
**Senior Editor** Sreshtha Bhattacharya
**Senior Art Editors** Anjana Nair, Chhaya Sajwan
**Editor** Suparna Sengupta
**Art Editors** Supriya Mahajan, Rakesh Khundongbam,
Pallavi Narain
**Assistant Art Editors** Ankita Mukherjee,
Namita, Shruti Singhal
**Managing Editor** Pakshalika Jayaprakash
**Managing Art Editor** Arunesh Talapatra
**Production Manager** Pankaj Sharma
**Pre-production Manager** Balwant Singh
**DTP Designers** Vishal Bhatia, Rajesh Singh Adhikari,
Nand Kishor Acharya, Syed Md Farhan
**Picture Researcher** Deepak Negi
**Picture Research Manager** Taiyaba Khatoon

This edition published 2016
First published in Great Britain in 2014
by Dorling Kindersley Limited,
80 Strand, London, WC2R 0RL

A CIP catalogue record for this book
is available from the British Library

ISBN: 978-1-4093-4792-7

Printed and bound in China

A WORLD OF IDEAS:
SEE ALL THERE IS TO KNOW

www.dk.com

# Contents

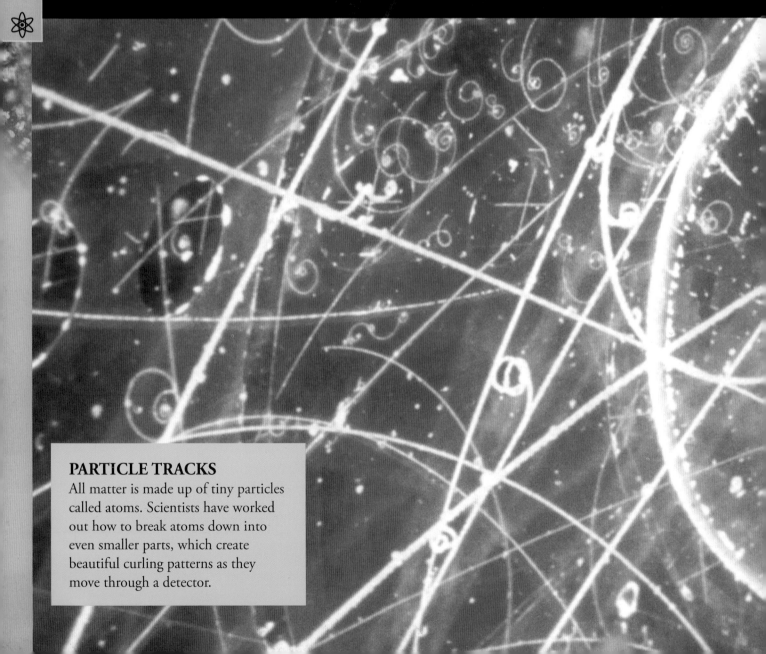

# MATTER

## PARTICLE TRACKS

All matter is made up of tiny particles called atoms. Scientists have worked out how to break atoms down into even smaller parts, which create beautiful curling patterns as they move through a detector.

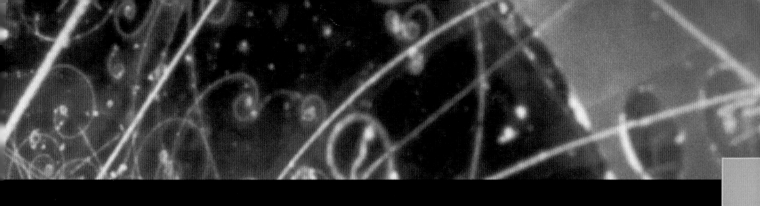

Matter is the stuff of the physical universe. You are made of matter. So is this book, the air around us, planet Earth, the Sun, and distant stars.

# Defining matter

Matter is the stuff of the Universe – solids, liquids, and gases. And matter is not just here on Earth. It makes up all the stars, planets, moons, and the bits of dust and wisps of gas drifting in space, as well as the Earth and everything on it. The whole Universe, as far as we know, is made up of matter and energy.

**STAR DUST**
Stars are made of matter. Sometimes, at the end of their lives, they explode, leaving behind clouds of dust called nebulae. Over billions of years, these nebulae come together and form new stars and planets. In fact our Earth, and everything on it, is made of star dust.

**STATES OF MATTER**
All matter on Earth exists in one of three forms: solid, liquid, or gas. Some matter is solid – it is hard, tough, and keeps its shape. Some is liquid – it flows or runs, but it cannot be squeezed smaller or stretched bigger. Some is gas – it flows, and it can expand or be squeezed. These three forms are known as the states of matter.

▲ GASEOUS *The particles of matter in a gas, such as water vapour, are far apart, move fast, and change their distance from each other.*

▲ SOLID *In a solid, such as ice, particles of matter are close together and fixed in place.*

▶ LIQUID *In water and other liquids, particles of matter are quite close, and can move around, but cannot change distance from each other.*

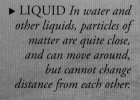

▲ SPARKLING STONES *This butterfly brooch is made from platinum and diamonds – two of the most expensive kinds of matter.*

## PRECIOUS MATTER

Some matter and materials are common and not very special, such as ordinary rocks and soil. Other matter and materials are valuable for various reasons. Diamonds, rubies, and other jewels are prized because they are rare, have beautiful colours, and can be polished to a bright shine.

## WOW!

Matter can be converted to energy. A tiny amount of matter makes an enormous amount of energy!

▼ STATES OF WATER *All you can see in this geyser at Yellowstone National Park, USA, is matter – steam (gas) rises from a pool of hot water (liquid), which is surrounded by snow (solid).*

## MATTER AND ENERGY

Matter can trap energy inside it. The gunpowder in fireworks contains lots of chemical energy. When it catches fire, it burns very fast, releasing the energy as light, heat, and sound. Petrol in a car does the same, providing energy to move us around.

▲ RELEASING ENERGY *Adding various substances to the gunpowder in fireworks makes them burn with different colours. Iron produces yellow sparks, and copper gives blue-green sparks.*

## OTHER FORMS OF MATTER

There are two states of matter that are only found under special conditions. Plasma is formed when the particles in a gas gain an electronic charge. It is found in the Sun, and also inside neon lights and plasma lamps. The other special state is called a Bose-Einstein condensate, which can be formed by cooling special gases to extremely low temperatures.

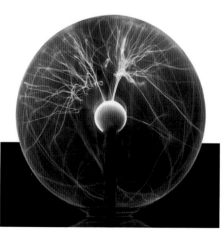

▶ PLASMA GLOBE *This globe contains a mixture of gases. A high-voltage electrode at the centre gives the particles of gas an electric charge, forming glowing plasma.*

# Atoms

All matter is made up of tiny particles known as atoms. For a long time, people thought that there was nothing smaller than an atom, but we now know that atoms in turn are made up of even smaller, "subatomic" particles called protons, electrons, and neutrons. Different types of atoms have different numbers of subatomic particles inside them. For example, hydrogen atoms have only one proton inside, while gold atoms have 197 protons.

### INSIDE AN ATOM

In the middle of an atom is the nucleus. This contains parts known as protons and neutrons, which hardly move from their central position. Whizzing fast around the nucleus are much tinier pieces called electrons. They travel round and round at a fixed distance from the nucleus, called a shell. Electrons only leave their shells if they absorb a jolt of energy.

## OLDER MODELS OF THE ATOM

We now know that atoms have a nucleus in the centre surrounded by electrons. In the past, scientists had different ideas about how particles inside an atom fit together, and came up with different models to describe what atoms look like.

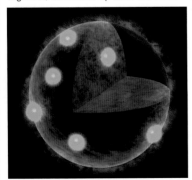

▲ "PLUM PUDDING" MODEL *In this old model, the parts of the atom are placed at random, like plums in a pudding.*

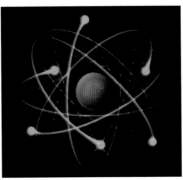

▲ PLANETARY MODEL *This later model shows electrons orbiting the nucleus of the atom, like planets around the Sun.*

**WOW!** Atoms are incredibly small. The dot on this "i" contains about one million, million atoms.

## PARTS OF AN ATOM

Despite their tiny size, each part of an atom has a mass (weight) and a type of electricity, or charge. A proton has a mass of 1 and a positive charge. A neutron also has a mass of 1 but carries no charge. An electron's mass is 1,850 times less than a proton and it has a negative charge. Since positive and negative charges attract, electrons keep moving around protons and do not fly away.

| PARTICLE | CHARGE | MASS | LOCATION |
|---|---|---|---|
| Proton | Positive (+) | 1 | Nucleus |
| Neutron | Neutral (0) | 1 | Nucleus |
| Electron | Negative (–) | 1/1,850th | Shell |

## MAKING MATTER

The number of protons, neutrons, and electrons in an atom decides what kind of substance it is. All atoms of sodium (a soft, silvery metal) have 11 protons, 12 neutrons, and 11 electrons. If the number of protons and electrons changes, a different substance is formed.

*Nucleus made up of protons (red) and neutrons (blue)*

*Inner layer ("shell") of electrons*

*Outer layer ("shell") of electrons*

▶ SODIUM LAMP *This lamp passes energy into sodium atoms, making them rush around and give off light. Different types of atoms give off different colours of light when they are energized.*

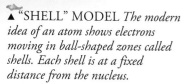

▲ "SHELL" MODEL *The modern idea of an atom shows electrons moving in ball-shaped zones called shells. Each shell is at a fixed distance from the nucleus.*

## SEEING ATOMS

The most powerful microscopes make things appear millions of times larger, and can even show atoms. These microscopes do not use light rays. They are electron microscopes – they use beams of electrons aimed very carefully to pick up tiny details.

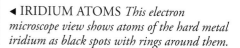

◀ IRIDIUM ATOMS *This electron microscope view shows atoms of the hard metal iridium as black spots with rings around them.*

# Molecules

Atoms rarely exist on their own. Usually they join with, or bond to, other atoms. Two or more atoms joined together are known as a molecule. If the atoms in a molecule are of different elements, they form a compound. Molecules vary hugely in size. Some are just two atoms, such as molecules of the gas oxygen in air. Other molecules have millions of atoms, in materials such as wood, plastic, and rubber.

## SALT CRYSTAL

In salt, each molecule has two atoms. One is sodium (Na), which when pure is a very lightweight metal. The other is chlorine (Cl), which on its own is a greenish poison gas. Joined together they produce a very different substance – sodium chloride (NaCl) – as tiny grains or crystals.

▶ TRANSLUCENT CRYSTAL *Billions of salt molecules arrange in regular patterns, like tiny bricks, to form a pyramid shape, known as a crystal. It has flat sides, angled edges, and sharp corners.*

## HOW ATOMS JOIN

Atoms join into molecules in several different ways. One is to "share" the outermost parts of their atoms, called electrons. An atom of sodium has just one electron in its outermost zone or shell. Chlorine has seven, with a neat space for just one more. So sodium's outermost electron spends some time in its own atom, and some in the chlorine atom. This keeps the two atoms near each other.

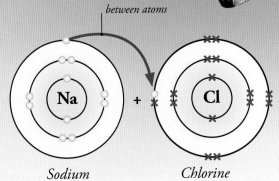

*Electron shared between atoms*

Na + Cl

*Sodium*          *Chlorine*

◀ SODIUM CHLORIDE *Electrons are negative. So the chlorine atom with an extra electron is negative, while the sodium atom without one is positive. Negative and positive attract, or pull nearer, helping the atoms stay together.*

# WOW!

In a human body, one DNA molecule can be 8 cm (3 in) – as long as your finger!

## SIMPLE MOLECULES

Most of the substances around us are compounds, which means their molecules contain atoms of more than one element. The simplest molecules have just two atoms, but even so they can be very different from the elements that make them up.

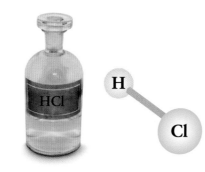

### Hydrochloric acid
*Each molecule has just two atoms – hydrogen (H) and chlorine (Cl) – making hydrochloric acid (HCl). This acid is very powerful. The human body produces it inside the stomach to break down and digest food.*

### Baking soda
*Used in cooking, cleaning, and medicine, this molecule has six atoms: one sodium (Na), one hydrogen (H), one carbon (C), and three oxygen (O). As $NaHCO_3$, it is usually known as sodium bicarbonate.*

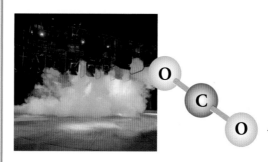

### Dry ice
*Dry ice (not frozen water) molecules have three atoms: two oxygen (O) and one carbon (C), forming carbon dioxide ($CO_2$). It makes misty, foggy effects for stage shows.*

### Chalk
*One atom of the metal calcium (Ca) joins to one carbon (C), and three oxygen (O) to make calcium carbonate ($CaCO_3$). One form of this is the bright white rock called chalk.*

## COMPLEX MOLECULES

Some molecules can contain hundreds, even thousands of atoms to make complicated structures. Carbon is particularly good at making complex molecules, as each atom of carbon can join to up to four neighbouring atoms. Most living things are made of molecules containing carbon.

◄ COFFEE BEANS *In beans, caffeine is mixed with more than 1,000 other substances, including lactones and phenylidanes, which cause the bitter taste.*

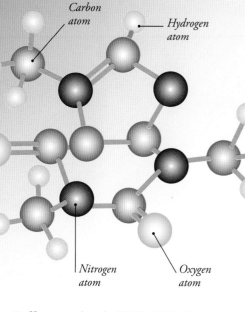

Carbon atom

Hydrogen atom

Nitrogen atom

Oxygen atom

**Caffeine molecule ($C_8H_{10}N_4O_2$)**

13

# Solids

In a solid, such as a block of brick or a lump of metal, the atoms or molecules are usually fixed in place. They cannot move around, or move nearer to or further from each other. This means most solid substances have a fixed shape that is difficult to change, except by powerfully squeezing or stretching them, or by breaking them apart. However, some solids are elastic, meaning their atoms can move apart slightly and come back together without breaking to pieces.

## INSIDE A SOLID

The molecules and atoms in a solid are joined by links, known as bonds. These bonds are very strong and difficult to bend or break, which is why the solid keeps its shape. The bonds hold atoms together in regular shapes such as rows of six-sided boxes.

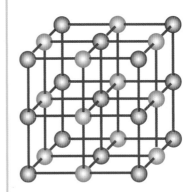

In this solid the molecules form rigid, stiff cubes.

## TYPES OF SOLID

Solids can be light or heavy, hard or soft, shiny or dull, sharp or smooth. Some, like see-through crystals, have no colour at all. The weight of a solid depends on the types of atoms inside it and the distance between them. Very heavy solids have close-packed, big, heavy atoms. The stronger the links, or bonds, between them, the more difficult it is to change the solid's shape.

**WOW!**
The heaviest naturally occurring solid is the metal osmium. It is twice as dense as lead.

▲ LIGHT *One of the lightest solids is graphene aerogel – it can balance on a flower. It is made of atoms of the substance carbon, with lots of empty space between.*

▲ MEDIUM *Wood has several kinds of small, light atoms, mostly of the substance carbon. There are also spaces between the atoms that were filled with water when the tree was alive.*

▲ HEAVY *Rocks such as basalt have atoms packed closely together, which are extremely difficult to move. The atoms, which include those of the metal iron, are also heavy.*

*Amethyst quartz contains atoms of iron, which give it its violet colour*

## CHANGING SHAPE

If you press or pull almost any solid with enough power, it will change shape or even break. While the molecules in the solid cannot move closer together, you can squeeze out air trapped in between. Large lumps of solid are hard to squash or stretch, but thin bars and sheets can often be bent or moulded. This is known as deforming a solid.

► CRUSHED *Cars have lots of thin parts, and an industrial crusher has enough pressure to squash them almost flat.*

## CRYSTAL SOLIDS

Some solids form crystals. These have flat faces, straight edges, and sharp corners. A crystal's shape is based on the way the different atoms and molecules inside fit against each other, like clipping together different-shaped building blocks. With more blocks, the shape gradually grows bigger, but keeps the same faces, edges, and corners.

◄ QUARTZ CRYSTAL *Quartz, such as this amethyst crystal, contains atoms of silicon and oxygen, which fit together into six-sided crystals that grow on each other. Tiny amounts of metal atoms give the quartz different colours.*

## FIBROUS SOLIDS

It is difficult to change the shape of a thick bundle of rope. Unwind the bundle and a single strand bends easily. Fibrous solids use a lot of fragile fibres twisted or woven together to create a much stronger fabric.

▼ FIBRES *Each fibre of rope or string is as thin as hair. Twisted together they are less bendy and much stronger.*

## CRYSTAL CAVE

Deep beneath the hills of northern Mexico lies a cave of natural wonders, full of giant white crystals of a mineral called gypsum. Intense heat and humidity from below the Earth's surface have made the crystals grow to enormous sizes – up to 12 m (39 ft) long and weighing 50 tonnes (55 tons).

# Liquids

In liquids, unlike in solids, the atoms and molecules can move around. But they cannot alter their distances from each other. This means a liquid can move and change shape, flowing to fill the space around it. It cannot change its volume – the amount of space it takes up – so liquids resist being squeezed or compressed.

## INSIDE A LIQUID

Atoms and molecules in a liquid are linked to each other, but in a loose, ever-changing way. They are free to move around in any direction, provided they stay the same distance from each other. They flow to the bottom of any container.

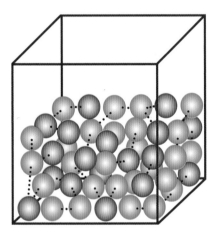

**In liquids, bonds break and reform, allowing atoms and molecules to move.**

## TWO LIQUIDS IN ONE

Milk is actually two liquids that do not mix, one floating about in the other. One liquid is tiny blobs, too small to see, of a substance called butterfat. These blobs are spread through the other liquid, which is mainly water. The scientific name for this type of double-liquid is an emulsion. Some kinds of paints are also emulsions.

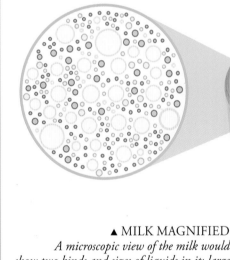

▲ MILK MAGNIFIED
*A microscopic view of the milk would show two kinds and sizes of liquids in it: large drops of butter fat and tiny molecules of water.*

## RESISTING PRESSURE

When trying to squeeze a liquid, the pressing force, called hydraulic pressure, spreads equally through all the liquid, which tries to escape through any gap or weak point. Turn on a tap and the water inside, under hydraulic pressure, quickly flows out.

◄ PRESSURE POWER
*Opening a fire hydrant lets the high-pressure water inside come out fast. The higher the pressure, the stronger the spray.*

## THICK AND THIN

Some liquids, such as water and petrol fuel, are runny and thin. They flow and spread easily. Others, such as treacle and diesel fuel, are thicker and gooey, and do not flow so well. This feature of liquids is known as viscosity. Low viscosity means more runny, while high viscosity means thick and sticky.

*Larger blobs of mercury form flat-topped pools with a shiny, mirror-like surface*

◄ READY TO ROLL

*Bitumen is a very thick or viscous liquid, made from crude oil (petroleum) from under the ground. It is used to lay road surfaces.*

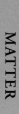

*Poured ribbons and folds of bitumen take a long time to spread out*

## LIQUID METAL

Most metals are hard solids at normal temperatures. The exception is mercury, which is a shiny, silvery liquid. It does not turn solid, or freeze, until it gets below –39°C (–38.2°F), which is twice as cold as inside a household freezer. (Water turns solid, or freezes, at 0°C or 32°F.)

### WOW!

Pitch, a hard black tar substance, is so viscous that when poured it takes 10 years between drops!

19

# Gases galore

Like a liquid, a gas has no shape of its own. But unlike a liquid, a gas can change its volume because its atoms and molecules can vary their distance apart. So a gas gets bigger and expands to fill its container, as its atoms and molecules spread out. If there is no container, the atoms and molecules keep spreading apart, further and further.

## GASES AND PRESSURE

In the same way that a gas expands to fill the space available, it can also be squeezed into a smaller space by a pushing force, or pressure. For example, when you pump air into a bicycle tyre, you are squeezing gas into a tight space. The atoms and molecules bump into each other more, which makes the gas hotter. Take away the pressure and the gas expands once again.

### INSIDE A GAS

Atoms and molecules in a gas are not linked to each other. They are free to wander, knock into each other, and bump into the sides of their container. Each time they collide, the atoms and molecules change direction but keep moving.

**In a gas the molecules roam freely.**

◀ FIZZING
*When the gas under pressure in this can is released, its tiny bubbles get bigger and make the drink fizz.*

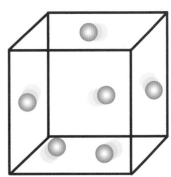

## WOW!

Tungsten hexafluoride is such a heavy gas that a balloon filled with it falls through the air, almost like a stone.

▲ HELIUM AIRSHIP *This huge airship is a giant version of a party balloon, filled with the very light gas helium.*

## FLOATING IN AIR

The air around us is a mixture of about 15 gases, each one weighing a different amount. The second-lightest gas is helium, making up five parts in every million parts of air. Collect enough helium to fill a balloon, even an airship, and it floats easily because it is lighter than air – that is, the mixture of gases all around is heavier.

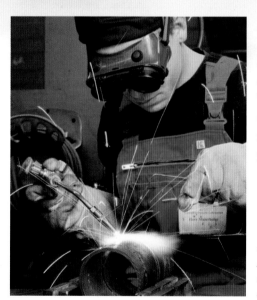

## BURNING GASES

Materials such as metals are joined by a method called welding. A gas called acetylene burns in oxygen to create a flame so hot that it melts metal. The white flame has a temperature of up to 3,500°C (6,330°F) – three times hotter than a log fire. The flame melts the edges of two pieces of metal so they flow together and, when cool, become one piece.

◄ HOT SMOKE
*A welder needs thick gloves and a pair of goggles to protect against the heat and brightness of a welding flame.*

## SLEEPING GAS

Every few seconds we breathe in air because we need one gas in it – oxygen – to keep us alive. Before a hospital operation, doctors may give patients a mixture of other gases, called an anaesthetic. This makes the brain go to sleep, so that the patient does not feel pain during the operation.

# Changing states

Most matter exists in one of three main states – solid, liquid, or gas. Things change between the three depending on their temperature. Rock is usually hard and solid. But when it is heated to above 1,000°C (1,832°F) deep underneath the Earth's crust, it melts to form a thick liquid known as magma. Water is a liquid at room temperature, but if we heat it to 100°C (212°F) it becomes a gas – water vapour.

## SOLIDS, LIQUIDS, AND GASES

Almost any solid substance will melt – change from solid to liquid – if it gets hot enough. Add more heat and it bubbles and boils, turning from liquid into gas. Take heat away from it and the reverse happens: it condenses from gas to liquid, then freezes or solidifies from liquid to solid.

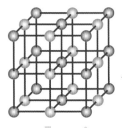

◄ SOLIDS
*Low-energy atoms and molecules stay fixed in position and hardly move.*

Melting  Freezing

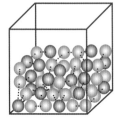

◄ LIQUIDS
*Medium-energy atoms and molecules flow but remain the same distance apart.*

Boiling  Condensing

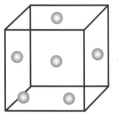

◄ GASES *High-energy atoms and molecules fly around at random, changing distance and direction.*

## CHANGING ENERGY

When you heat a solid substance, its atoms get more energy. They begin to move about more, or vibrate. Eventually the atoms have enough energy to break out of their fixed positions in the solid. Then they start to move around and flow – at this stage the solid becomes a liquid.

▲ MELTING CHOCOLATE *As heat seeps into solid chocolate from the outside, the parts around the edge melt. The inner parts are still slightly cooler, so they stay solid for a little longer.*

| SUBSTANCE | MELTING POINT | BOILING POINT |
| --- | --- | --- |
| Water | 0°C (32°F) | 100°C (212°F) |
| Chocolate | 30–35°C (86–95°F) | 110–120°C (230–248°F) |
| Candle wax | 60–65°C (140–149°F) | 240–250°C (464–482°F) |
| Cooking oil | –17°C (–63°F) | 225–230°C (437–446°F) |
| Lead | 327°C (620.6°F) | 1,749°C (3,180.2°F) |

## MELTING AND BOILING POINT

Every substance changes state at particular temperatures, known as the melting and boiling points. Pure chemicals, such as water or lead, have very precise melting and boiling points. Substances such as chocolate and cooking oil contain different ingredients mixed together, so their melting and boiling points can change.

### WOW!

The metal rhenium has the highest boiling point of any element. It turns to gas at 5,596°C (10,105°F).

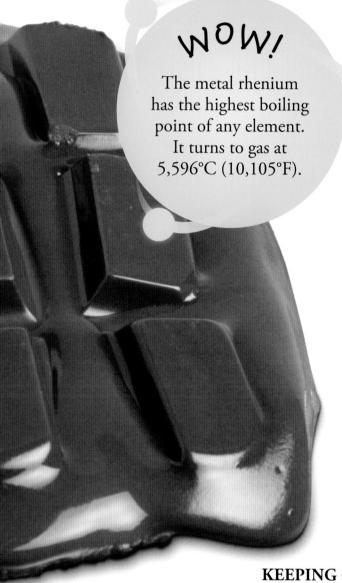

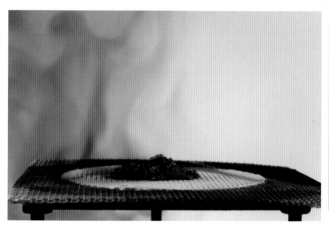

## MISSING OUT A STAGE

Some chemicals do not melt with increasing temperature. Instead, they turn from solid directly to gas, in a process called sublimation. These chemicals include carbon dioxide, iodine (above), and arsenic. Under suitable conditions, water ice can also turn directly into water vapour. The reverse process, when a gas changes directly into a solid, is called deposition.

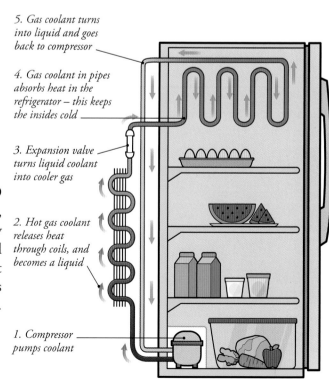

5. Gas coolant turns into liquid and goes back to compressor

4. Gas coolant in pipes absorbs heat in the refrigerator – this keeps the insides cold

3. Expansion valve turns liquid coolant into cooler gas

2. Hot gas coolant releases heat through coils, and becomes a liquid

1. Compressor pumps coolant

### KEEPING COLD

Changing states involves energy. For example, melting takes in heat energy, while condensing gives energy out. A refrigerator works by turning a substance, called coolant, from liquid to gas and back again. Liquid coolant boils as it takes heat from inside the refrigerator, then cools to a liquid again by passing the heat to the air outside.

▶ ELECTRIC COOLING *A refrigerator needs the energy of electricity to work its compressor. This sends the coolant round and round, moving heat from inside to outside.*

# All the elements

An element is a pure chemical substance. All of its atoms are the same, and they are different from the atoms of any other substance. There are more than 100 elements. Some we see and use every day, while others are very rare and have strange features. Each element has a scientific symbol of one or two letters, such as H for hydrogen and Ca for calcium.

## SEEN AND UNSEEN

The periodic table allows us to predict how one element will react with another. The element iron, with the scientific symbol Fe, is a hard and shiny solid. In a common chemical reaction, one atom of iron joins with two atoms of the element oxygen (O) from the air all around us. The result is the red-brown powder known as iron oxide or rust ($FeO_2$).

**Iron nail**

**Rusty nail**

▲ TWO ELEMENTS JOINED *Shiny iron and invisible oxygen link together to form a very different substance – the compound iron oxide (seen as rust).*

## REACT OR NOT?

The elements at the far left of the table, alkali metals, are rarely on their own in nature. They are too reactive – they nearly always join with other elements to make compounds. On the right of the table in the final column of elements, noble gases are the most unreactive – they hardly ever join with other atoms.

## TABLE OF ELEMENTS

Elements are listed in a chart known as the periodic table. It shows two main features. Most of the elements on the left are metals, while those on the right are non-metals. The elements in the rows, or periods, across the top are lighter, while towards the bottom, they get heavier.

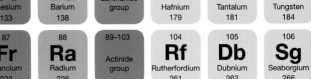

▲ SODIUM *A very lightweight metal, sodium always wants to react, even with water or air – it can suddenly burst into flames.*

▲ XENON *A very rare gas, xenon is very unreactive. It has a few special uses in electronics, lasers, and medical scanners.*

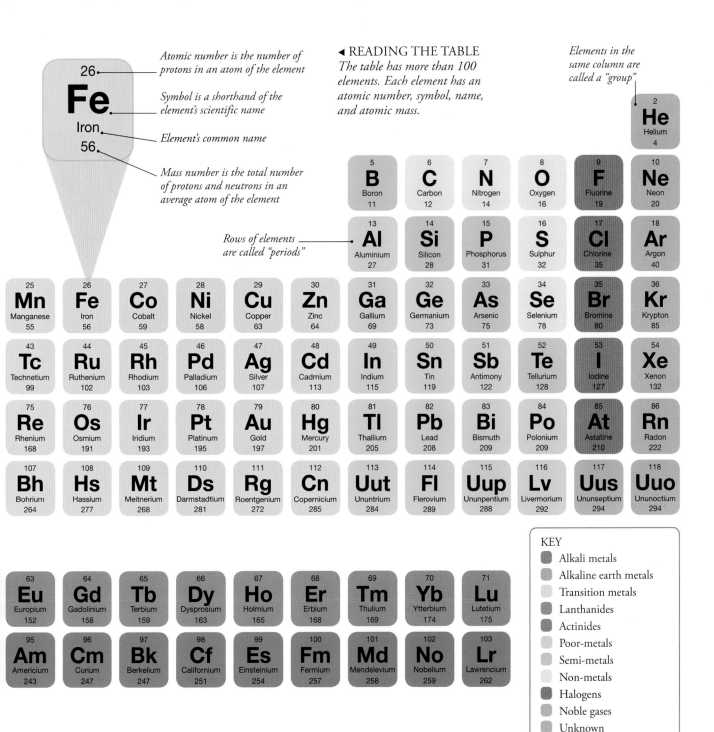

**26**
**Fe**
Iron
**56**

Atomic number is the number of protons in an atom of the element

Symbol is a shorthand of the element's scientific name

Element's common name

Mass number is the total number of protons and neutrons in an average atom of the element

◄ READING THE TABLE
*The table has more than 100 elements. Each element has an atomic number, symbol, name, and atomic mass.*

*Elements in the same column are called a "group"*

*Rows of elements are called "periods"*

| | | | | | | | | | | | | 2<br>**He**<br>Helium<br>4 |

| 5<br>**B**<br>Boron<br>11 | 6<br>**C**<br>Carbon<br>12 | 7<br>**N**<br>Nitrogen<br>14 | 8<br>**O**<br>Oxygen<br>16 | 9<br>**F**<br>Fluorine<br>19 | 10<br>**Ne**<br>Neon<br>20 |

| 13<br>**Al**<br>Aluminium<br>27 | 14<br>**Si**<br>Silicon<br>28 | 15<br>**P**<br>Phosphorus<br>31 | 16<br>**S**<br>Sulphur<br>32 | 17<br>**Cl**<br>Chlorine<br>35 | 18<br>**Ar**<br>Argon<br>40 |

| 25<br>**Mn**<br>Manganese<br>55 | 26<br>**Fe**<br>Iron<br>56 | 27<br>**Co**<br>Cobalt<br>59 | 28<br>**Ni**<br>Nickel<br>58 | 29<br>**Cu**<br>Copper<br>63 | 30<br>**Zn**<br>Zinc<br>64 | 31<br>**Ga**<br>Gallium<br>69 | 32<br>**Ge**<br>Germanium<br>73 | 33<br>**As**<br>Arsenic<br>75 | 34<br>**Se**<br>Selenium<br>78 | 35<br>**Br**<br>Bromine<br>80 | 36<br>**Kr**<br>Krypton<br>85 |

| 43<br>**Tc**<br>Technetium<br>99 | 44<br>**Ru**<br>Ruthenium<br>102 | 45<br>**Rh**<br>Rhodium<br>103 | 46<br>**Pd**<br>Palladium<br>106 | 47<br>**Ag**<br>Silver<br>107 | 48<br>**Cd**<br>Cadmium<br>113 | 49<br>**In**<br>Indium<br>115 | 50<br>**Sn**<br>Tin<br>119 | 51<br>**Sb**<br>Antimony<br>122 | 52<br>**Te**<br>Tellurium<br>128 | 53<br>**I**<br>Iodine<br>127 | 54<br>**Xe**<br>Xenon<br>132 |

| 75<br>**Re**<br>Rhenium<br>168 | 76<br>**Os**<br>Osmium<br>191 | 77<br>**Ir**<br>Iridium<br>193 | 78<br>**Pt**<br>Platinum<br>195 | 79<br>**Au**<br>Gold<br>197 | 80<br>**Hg**<br>Mercury<br>201 | 81<br>**Tl**<br>Thallium<br>205 | 82<br>**Pb**<br>Lead<br>208 | 83<br>**Bi**<br>Bismuth<br>209 | 84<br>**Po**<br>Polonium<br>209 | 85<br>**At**<br>Astatine<br>210 | 86<br>**Rn**<br>Radon<br>222 |

| 107<br>**Bh**<br>Bohrium<br>264 | 108<br>**Hs**<br>Hassium<br>277 | 109<br>**Mt**<br>Meitnerium<br>268 | 110<br>**Ds**<br>Darmstadtium<br>281 | 111<br>**Rg**<br>Roentgenium<br>272 | 112<br>**Cn**<br>Copernicium<br>285 | 113<br>**Uut**<br>Ununtrium<br>284 | 114<br>**Fl**<br>Flerovium<br>289 | 115<br>**Uup**<br>Ununpentium<br>288 | 116<br>**Lv**<br>Livermorium<br>292 | 117<br>**Uus**<br>Ununseptium<br>294 | 118<br>**Uuo**<br>Ununoctium<br>294 |

| 63<br>**Eu**<br>Europium<br>152 | 64<br>**Gd**<br>Gadolinium<br>158 | 65<br>**Tb**<br>Terbium<br>159 | 66<br>**Dy**<br>Dysprosium<br>163 | 67<br>**Ho**<br>Holmium<br>165 | 68<br>**Er**<br>Erbium<br>168 | 69<br>**Tm**<br>Thulium<br>169 | 70<br>**Yb**<br>Ytterbium<br>174 | 71<br>**Lu**<br>Lutetium<br>175 |

| 95<br>**Am**<br>Americium<br>243 | 96<br>**Cm**<br>Curium<br>247 | 97<br>**Bk**<br>Berkelium<br>247 | 98<br>**Cf**<br>Californium<br>251 | 99<br>**Es**<br>Einsteinium<br>254 | 100<br>**Fm**<br>Fermium<br>257 | 101<br>**Md**<br>Mendelevium<br>258 | 102<br>**No**<br>Nobelium<br>259 | 103<br>**Lr**<br>Lawrencium<br>262 |

**KEY**
- Alkali metals
- Alkaline earth metals
- Transition metals
- Lanthanides
- Actinides
- Poor-metals
- Semi-metals
- Non-metals
- Halogens
- Noble gases
- Unknown

## STRANGE ELEMENTS

A few elements are so common that we mention their names often, such as carbon, iron, and tin. Others are rare and strange, with unfamiliar names and few uses. Some elements take their names from famous scientists, from the places the elements were first discovered or purified, or from ancient languages. The symbol for lead, Pb, comes from the Ancient Roman name for lead, *plumbum*.

39
**Y**
Yttrium
89

### Yttrium

- ■ **Property** Soft, silvery metal
- ■ **Atomic number** 39
- ■ **Atomic mass** 89

This metal is named after the village of Ytterby in Sweden, where rocks rich in Yttrium were collected for chemical tests. It was first made in pure form in 1828.

87
**Fr**
Francium
223

### Francium

- ■ **Property** Heavy, reactive metal
- ■ **Atomic number** 87
- ■ **Atomic mass** 223

Discovered in 1939, francium is very hard to purify. It tries to combine with other elements and its atoms rapidly break apart, giving off radioactivity.

# Mixtures

A mixture is different substances mixed together, but no chemical reactions have happened between them. They may be swirled up and jumbled together, but all the individual parts are still on their own. They can be separated again fairly easily. Mixtures of big things, such as rocks and sand, are easy to spot, but other mixtures, such as the air around us, are harder to see, and harder to split apart.

## WOW!

Seawater is a mixture of more than 100 different substances including salts, minerals, ores, and gases.

## MIXTURES ALL AROUND

A rocky shore is a mixture with all kinds and sizes of materials – sand, shingle, pebbles, and boulders heaped together. But waves may help to separate these parts into different sizes, known as sorting. Slow-moving water easily picks up small grains of sand and swishes them together. Shingle is bigger and heavier and needs stronger waves, while pebbles are heavier still. Big boulders are hardly ever moved, even by the most powerful waves, and so are left higher on the shore.

*Air is a mixture of several gases*

*Seawater contains mixtures dissolved from rocks*

◀ BIGGEST *Rocks are mixtures of several different chemicals.*

◀ SMALLEST *Sand is a mixture of many kinds of rock and shell, all ground up together.*

## DIFFERENT KINDS OF MIXTURE

Some mixtures have large or coarse particles that are easy to see. Some have tiny ones we can hardly see, which float and stay still, or suspend, in a liquid. An alloy is a mix of a metal with another substance. For example, steel is made of iron mixed with other metals and carbon to make it stronger. In a solution, the tiniest particles of a substance float in a liquid, but can be separated by boiling the liquid away.

# SEPARATION

A mixture can usually be divided into its separate parts physically, for example by shaking, filtering, or floating. This is because the parts have not joined or combined chemically to produce new substances, as they do in molecules and compounds. When a mixture's parts are sorted out and separated from each other, they are the same as before they were mixed together.

Cork floats on oil. Oil floats on water

Plastic brick sinks in oil but floats on water

Onion sinks through oil and water but floats on syrup

Syrup sinks below water

◄ FLOATING *Shaken together, these substances mix. Left for a time in a jar, they separate again. The lightest ones float to the top – liquid oil, as well as the solid cork.*

▲ MAGNETIZING *Iron is pulled towards, or attracted by, a magnet, while sand is not. So a magnet can attract tiny bits of iron from a mix with sand, leaving behind the sand grains.*

▼ FILTERING *A filter, strainer, or sieve has holes that allow small items through but hold back larger ones. A common example is catching tea leaves in a strainer.*

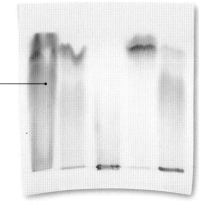

Tea leaves trapped by mesh of strainer

Water flows through strainer

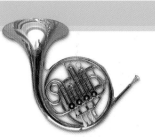

Different substances travel across the paper at different speeds, separating them from each other. This can help us see what is in the mixture.

## SEPARATED BY SIZE AND SPEED

In the scientific method called chromatography, a liquid or gas is added to a mixture. This is then passed through a special substance, such as paper or gel. The smallest particles of the mixture move along fastest, the medium ones slower, and so on. The result is that the particles of the mixture separate by size, so we can analyse the mixture.

▲ COARSE MIXTURE *The different parts are easy to see in a seed mix. Sand and gravel are also coarse mixtures.*

▲ SUSPENSION *In muddy water, tiny particles of soil float around (are suspended) in the liquid.*

▲ ALLOY *The alloy brass, used to make instruments, is a mixture of copper and zinc. It combines features of both metals.*

▲ SOLUTION *Substances such as sugar dissolve in water. If the water dries out, the sugar is left behind.*

# Solutions and solvents

Some liquids, such as water, can pull apart solids. You can see this if you add sugar to water – the sugar seems to disappear as its molecules are pulled in different directions by molecules of water. This is a special kind of mixture called a solution.

## SMALLER AND SMALLER

We may be able to see a solid substance breaking apart into smaller and smaller pieces in a process called dissolving. It may break into several bits. Each lump gets smaller as it loses material from the outside, which floats away in the liquid. The liquid in a solution is known as the solvent, while the stuff that dissolves is called the solute.

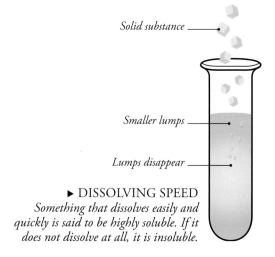

Solid substance

Smaller lumps

Lumps disappear

► DISSOLVING SPEED
*Something that dissolves easily and quickly is said to be highly soluble. If it does not dissolve at all, it is insoluble.*

## DILUTE AND CONCENTRATED

When a solute dissolves, it breaks up into tiny pieces – usually its smallest possible form, called molecules. These mix and float about among the molecules of the solvent. Just a little solute makes a dilute solution, while a lot makes it concentrated. The solute and the solvent do not join or combine. They form a mixture and can be separated again.

◄ FIZZING *As a tablet dissolves in water, some bits of it combine with water to make gas bubbles, while the rest dissolves and becomes invisible.*

## SATURATED SOLUTIONS

As more solute is added to a solution, the spaces between the solvent's molecules gradually fill. Eventually there is no room left, so no further solute can dissolve. Any more of it stays as a solid. The solution is then said to be saturated. Heating the solution increases the movement of the solvent molecules, allowing a bit more solute to dissolve.

◄ BACK TO SOLID *When a hot saturated solution cools, some of the solute can no longer stay dissolved. It turns back into a solid, usually crystals, like the sodium acetate in this picture.*

## COMMON SOLVENTS

The most common solvent is water, in which many substances dissolve. In daily life we use other solvents for special purposes – sometimes for solutes that do not dissolve in water, and therefore are hard to wash away.

| NAME OF SOLVENT | DESCRIPTION |
|---|---|
| White spirit | Makes paint thinner (more dilute), cleans paintbrushes, removes paint stains and grease |
| Nail varnish remover | Gets rid of nail varnish, which is specially designed not to be washed away by water |
| Laundry stain remover | A mix of solvents that dissolve various substances which water cannot dissolve, such as fat and wax |
| Hand cleanser | Breaks up, dissolves and removes oil and grease, plus the dirt and germs stuck in them |

## DISSOLVED COLOURS

Watercolour paints are made from special substances called pigments, which have strong colours and are soluble in water. Mix the paint with water and these coloured pigments dissolve and spread out. After painting, the water evaporates (dries) and leaves behind the coloured pigment. Other types of paints have different solvents, such as alcohol or oil.

## WOW!

The solvent DMSO (dimethyl sulphoxide) dissolves thousands of substances – which makes it very poisonous.

## NOT DISSOLVING

Oil does not mix with or dissolve in water. An oil spill in a river, lake, or sea floats on the water's surface, causing terrible pollution and harming wildlife. Chemicals called solvents and dispersants break the oil into tiny blobs, which spread out and float away, causing less harm.

# Acids and bases

Liquid substances have a level of acidity, a property that describes how they react to other substances. At the opposite ends of this scale, "acids" and "bases" are made by dissolving particular substances in water. At their strongest they can be very dangerous, eating away metals and other solids and causing burns if they touch human skin. In their weaker forms, acids and bases are safe to eat, and are very common in food and drink.

WOW!

"Magic acid" is a mixture of chemicals that is so acidic it can dissolve plastics.

## HOW STRONG IS IT?

Scientists use a scale called pH to measure how acidic or basic a liquid is. The scale runs from 1 for very strong acids to 14 for very strong bases. Neutral liquids have a pH of 7. We can measure pH with special chemicals called indicators, which change colour to show how acidic or basic a liquid is. The most common is universal indicator, which shows reds and orange for acids, greens for neutral, and blues and purples for basic solutions.

▲ VINEGAR Common, or table, vinegar is a solution of acetic acid and has a pH of 2–3.

▲ TOMATO JUICE Tomato juice contains citric acid, giving it a pH of around 5–6.

*Acid*

## ACIDS

Strong acids can eat away metals and some kinds of stone. They are used in the chemical industry, and in batteries for cars and other vehicles. Hydrochloric acid, one of the strongest, is created inside your stomach to break down the food you eat. Weak acids have a sharp flavour, which you can taste in many kinds of fruit and in vinegar.

▲ BATTERY ACID A vehicle battery contains very strong sulphuric acid with a pH of 1–2.

**THE pH SCALE**

| 0 | 1 | 2 | 3 | 4 | 5 | 6 | 7 | 8 | 9 | 10 | 11 | 12 | 13 | 14 |
|---|---|---|---|---|---|---|---|---|---|----|----|----|----|----|

Strong acid · Weak acid · Neutral · Weak base · Strong base

▲ ORANGE JUICE Citrus fruits such as lemon and grapefruit have acidic juice. Orange juice has a pH of 3–5.

## BASES

Strong bases can be useful cleaning materials, as they are good at dissolving grease and killing bacteria. They must be handled with care as they can cause burns to unprotected skin. Weaker bases are sometimes used in medicine to neutralize acids in the body, for example to treat stomach pain caused by excess acid.

## NEUTRALIZING

When acids and bases come into contact, they undergo a chemical reaction called "neutralization". The acid and base cancel each other out, to leave behind water and a kind of chemical called a salt. Sometimes a gas is also produced.

◄ BUBBLING OVER *Add red dye to baking soda inside a model volcano, drip on vinegar, and the volcano bubbles and "erupts"!*

▲WATER *Pure water is almost exactly neutral, with a pH very close to 7.*

▲ ANTACID TABLET *Taken to cancel out too much stomach acid, antacids are around pH 10.*

▲ DRAIN CLEANER *A chemical called lye is a very strong base, often used for cleaning. Its pH is 13–14.*

*Base*

▲ COW'S MILK *Most kinds of cow's milk are very slightly acidic, at pH 6.5–6.8.*

▶ TOOTHPASTE *Because of the substances added to strengthen teeth, toothpaste is a base with pH 8–9.*

▼ CEMENT *Cement includes a chemical called calcium hydroxide, which is a powerful base with pH 11–13.*

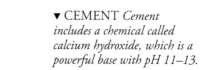

# Incredible reactions

Many substances stay the same for a long time. Some mountains have been around for billions of years. But other substances change – and not just by dissolving, or altering state from liquid to gas. Their molecules break apart into their separate atoms, and these join or bond together in new ways and new combinations. This is called a chemical reaction because the original substances react or alter each other, to make a new, different chemical substance.

MATTER

◄ SLOW AND STEADY *As soon as water is added to cement, chemicals in the cement dissolve and start to react. After a time, they set into a hard solid.*

## SLOW REACTIONS

Cement is a mixture of chemicals, which react together when water is added. The reaction may take a few hours or even days, but at the end of it the powdery cement has set into a hard solid. Slow reactions give out heat just like fast ones, but much more gradually.

## FAST REACTIONS

Some chemical reactions take place very quickly. Dynamite explodes in a fraction of a second. Adding an acid to a base can produce a violent reaction in just a second or two, with fizzing and bubbling as a gas is given off. Both reactions produce heat.

► SWIRLS AND FUMES *Adding two chemicals together must always be done under safe conditions. Here they react quickly, swirling the liquid and giving off gas fumes.*

---

### CHEMICAL REACTION

Iron (Fe)  +  Sulphur (S)  →  Iron sulphide (FeS)

The atoms in an element, such as iron or sulphur, are joined together by strong forces called bonds. When iron and sulphur are combined in a chemical reaction, they form iron sulphide. The properties of iron sulphide are different from its elements.

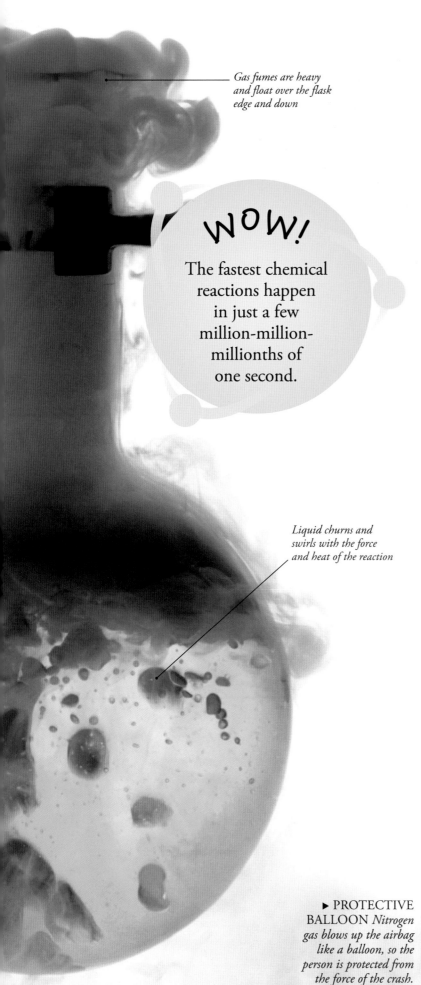

Gas fumes are heavy and float over the flask edge and down

## WOW!

The fastest chemical reactions happen in just a few million-million-millionths of one second.

Liquid churns and swirls with the force and heat of the reaction

▶ PROTECTIVE BALLOON *Nitrogen gas blows up the airbag like a balloon, so the person is protected from the force of the crash.*

## HOT REACTIONS

One of the most familiar chemical reactions is burning, also known as combustion. This is a reaction where a fuel, such as wood, coal, charcoal, or gas, combines with oxygen in the air. Unlike the other reactions shown on this page, combustion is not spontaneous – it does not start on its own. The fuel has to become very hot – that is, we light the fire. Once the reaction starts, it gives out its own heat, which lights more fuel.

▲ PRODUCING ENERGY *When charcoal burns, its atoms rearrange to form new molecules. This reaction gives off energy in the form of heat, light, and often sound.*

## LIFE-SAVING REACTIONS

Reactions are all around us, in cooking, cleaning, vehicle engines, and many other parts of daily life. One very fast reaction that helps to save lives is in a vehicle airbag. A sudden stop makes the chemical sodium azide ($NaN_3$) in the airbag split apart, or decompose, to give off nitrogen gas ($N_2$), which fills the airbag. This all happens in just 1/25th of a second.

## CHEMICAL REACTION

Some chemical reactions are slow and gentle, but others are fast and release a lot of energy. The chemicals in this picture are old military explosives which are being detonated to dispose of them safely. They react to produce heat energy and hot gas, which burst outwards in an explosion.

# Metals

About 90 of the pure substances, or chemical elements, are metals. They are used to make many familiar things around us, such as aluminium in fizzy-drinks cans, copper in electrical wires, iron in saucepans, nickel or lithium in rechargeable batteries, and solid gold in jewellery. Most metals are strong and tough. They are also good conductors, as they carry electricity and heat well.

## COMMON METALS AND THEIR USES

About 30 of the 90 metals are known as transition metals. These form the central block of the periodic table. Transition metals are typically hard and shiny. They can be hammered and bent into shape, and have high melting points. They are the best conductors, and do not react easily with other substances. These features make them very useful.

▲ GOLD *A symbol of wealth and power, gold polishes well and has a rich colour, which does not fade or rust. This mask of the Egyptian king Tutankhamun (reigned 1333–1323 BCE) is made of 10 kg (22 lb) of gold.*

*Titanium frame*

▲ TITANIUM *This lightweight metal has super strength, which is why it is used in some bicycles frames.*

▲ COPPER *It is often used in pipes and in electrical wiring.*

▶ STAINLESS STEEL *Steel is mostly made of iron. The stainless type has other substances mixed in to prevent rusting.*

## MAKING METALS

Some metals are found in a natural pure form, such as gold nuggets. But usually their metal atoms are combined with other elements in rocks. If a rock is rich in metal, it is known as an ore. The ore is heated with chemicals until it melts, in a process called smelting. The pure metal separates out as a liquid, which cools and hardens.

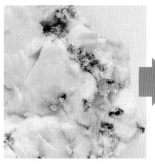

▲ 1. ORE *Sometimes small pieces of metal can be seen in the ore, such as these flecks of gold.*

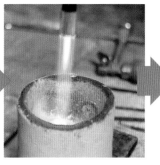

▲ 2. MELTING *The heat is carefully controlled to make the ore melt but not turn to vapour.*

▲ 3. COOLING *The melted liquid gold collects and can be poured or skimmed off to solidify.*

## WOW!

Tungsten, the hardest metal, also has the highest melting point – 3,422°C (6,192°F), or three times higher than gold.

### COLOURED METALS

Many strongly coloured substances, called pigments, contain transition metals. The colour does not come from the metal, but from compounds made when the metal reacts with other elements. Copper compounds tend to be blue-green or green, manganese compounds are brown, and cobalt compounds are an intense blue.

**Metallic coloured pigments**

## ANCIENT ALLOY

The Bronze Age began more than 5,000 years ago when people discovered that melting and mixing two metals – copper and tin – made a much harder material. The mixed metal, or alloy, was bronze. It made stronger, sharper swords, spears, and other weapons, as well as long-lasting bowls, dishes, vases, and other implements. Around 3,200 years ago another metal, iron, gradually took over from bronze.

*Rhodium-coated gold alloy ring*

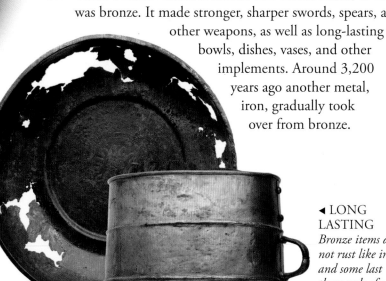

◄ LONG LASTING *Bronze items do not rust like iron, and some last thousands of years.*

## MORE VALUABLE THAN GOLD

The metal rhodium is rare, hard, shiny, and difficult to obtain, as well as very costly. It has a very high melting temperature of almost 2,000°C (3,632°F), much higher than gold (1,064°C or 2,919°F) and iron (1,538°C or 2,800°F). Depending on supply and demand around the world, rhodium can cost 10 times more than gold.

# Strange metals

The metals we see every day, such as iron, silver, and copper, come from a group called transition metals. They are hard, shiny, and only melt at high temperatures. There are other kinds of metals that behave very differently. They are divided into groups such as alkali metals, alkaline earth metals, and poor metals. Some are common in daily life while others are rare, with specialized uses, and a few are very dangerous.

*Lead weights*

## HEAVY AND SOFT

Lead, like aluminium, is a "poor metal". It is heavy, soft, and bendy, but resistant to rusting and crumbling. It used to be popular for roofing and for items that needed to be heavy but small, such as weights in old clocks. But lead was discovered to harm humans, other living things, and the environment, and so many of its uses have faded away.

## LIGHT YET STRONG

Aluminium comes from a group known as "poor metals". They are softer and easier to melt than transition metals. However, aluminium can be hardened by adding chemicals to make a strong, lightweight metal that is easy to shape and does not rust. This is why it is used in many items, from cooking foil and saucepans to ladders, casings for electronic gadgets, magnets, aircraft, and all kinds of machinery.

## WOW!

The metal caesium is so reactive that it catches fire in the air and, if added to water, it will cause an explosion.

▲ DULL SHINE *Silvery-coloured aluminium is often used for building aircraft, as it is strong, yet light and easily shaped into curved sheets.*

Pure silicon mixed with other elements is the main material used in computer chips

## SEMICONDUCTORS

The substances silicon and germanium are partly like metals and partly not, so they are called metalloids. They are also semiconductors, meaning they only conduct electricity under certain conditions. These in-between features make them ideal for use in microchips in electronic gadgets, such as mobile phones, and computers.

▲ SILICON POWER *Silicon microchips are the thinking power behind computers like this laptop.*

## ALKALINE EARTH METALS

Calcium is in the metal group called alkaline earths. When pure it is grey and soft. But it is hardly ever found pure since it reacts so easily with many other substances to form salts and other compounds, and some of these are very hard and tough. Calcium compounds can be found in the hard parts of living things, such as bones, teeth, and shells.

▲ SHELLS *About half of a seashell is calcium in the form of calcium carbonate. The shell lasts much longer than the animal inside.*

All kinds and sizes of planes use aluminium as it resists rusting

## USE WITH CARE

One of the most poisonous substances is arsenic. It is not quite a metal, nor a non-metal, and so is known as a metalloid. Arsenic affects the digestive system of human beings and animals. That is why it is used in controlled amounts in some items, such as insecticides. In nature it can soak through rocks into water supplies, making the water deadly to drink.

# Not at all metals

Of the more than 100 natural chemical elements, only about 17 are non-metals. Most of these are gases at room temperature, while others, such as sulphur, phosphorus, and carbon, are solids. They are mostly soft rather than hard, powdery rather than shiny, and do not carry heat or electricity.

## SULPHUR

Pure sulphur is a bright yellow powder made of tiny crystals. It sometimes occurs naturally in rocks and around hot springs. Sulphur is one of the most important substances in industry, used to make hundreds of different chemicals, rubbers, papers, fertilizers, pesticides, and preservatives.

WOW!

Jupiter's moon Io is covered in huge yellow plains made of the non-metal sulphur.

▼ SULPHUR SPRINGS
*At this hot spring in Dallol, Ethiopia, water containing natural sulphur and other minerals has dried up, leaving behind a yellow crust.*

## STRIKE A MATCH

Pure phosphorus is not found in nature because it reacts with other substances too easily, forming phosphates and similar compounds. These compounds are obtained from ore rocks and are widely used as fertilizers. Some forms of phosphorus catch fire easily when rubbed and are used in match heads.

## GERM KILLERS

One group of non-metals is known as halogens. These are fluorine, chlorine, bromine, iodine, and astatine. In pure form fluorine and chlorine are gases, bromine is a liquid, and iodine and astatine are solids. All halogens, when used in certain quantities, can be harmful, or toxic, to living things. Yet they can also be used with great care in health and medicine.

◄ FLUORINE *In its pure form, fluorine gas can damage the eyes, nose, airways, and inner organs. But tiny amounts in toothpaste help teeth to stay strong.*

◄ IODINE *Less toxic than other halogens, iodine and its compounds are helpful germ killers, as antiseptics on skin and disinfectants on non-living objects.*

◄ CHLORINE *A greenish choking gas at ordinary temperature, chlorine is added to water to kill germs, while larger amounts are used in bleaches.*

## NOBLE GASES

The six chemicals on the far right of the table of elements (p.24) rarely react with any substances, including themselves. They are helium, neon, argon, krypton, xenon, and radon. Because they are all gases and they were once seen as "better" than or "superior" to other chemicals, they are known as noble gases.

▶ NEON LIGHTS *Since noble gases are so unreactive, when electricity passes through neon glows without reacting, and so it makes long-lasting lights.*

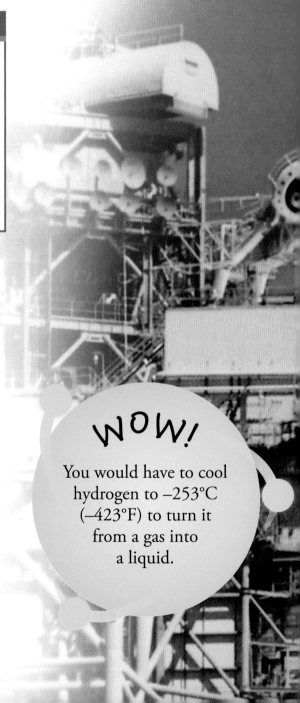

# Hydrogen

Hydrogen is the lightest of all the elements. Each of its atoms has just one electron moving around one proton. Hydrogen is about three-fourths of the mass of all the atoms we can see in the Universe. On Earth, it is one of the elements in water ($H_2O$), but as a gas it is only found in very small traces in our atmosphere. As hydrogen burns easily in air and creates almost no pollution, it could become a fuel of the future.

## HYDROGEN FACTS

- **Formula** $H_2$
- **Group** Non-metals
- **Density** 0.09 gm/l (0.012 oz/gallon)
- **Melting point** −259°C (−434°F)
- **Boiling point** −253°C (−423°F)
- **Main sources** Treating oil (petroleum) with super-hot steam; splitting water using electricity
- **Main uses** Many industries, for example, separating substances in oil; making strong acids and bases; altering foods such as margarines and cooking oils; welding; electronics; keeping things ultra-cold

Hydrogen molecule

## FLOATS AND BURNS

Hydrogen is about 14 times lighter than air. So a balloon filled with hydrogen floats upwards like a bubble in the heavier air around it. Many years ago, hydrogen was used in huge balloons and airships, lifting them up into the air. As hydrogen catches fire easily when mixed with air, and even explodes, airships today use safer gases.

▲ TIED DOWN *A balloon filled with hydrogen is kept from floating away by a heavy weight.*

▲ BLOWN UP *A flame melts the balloon, and sets fire to the hydrogen inside.*

## WOW!

You would have to cool hydrogen to −253°C (−423°F) to turn it from a gas into a liquid.

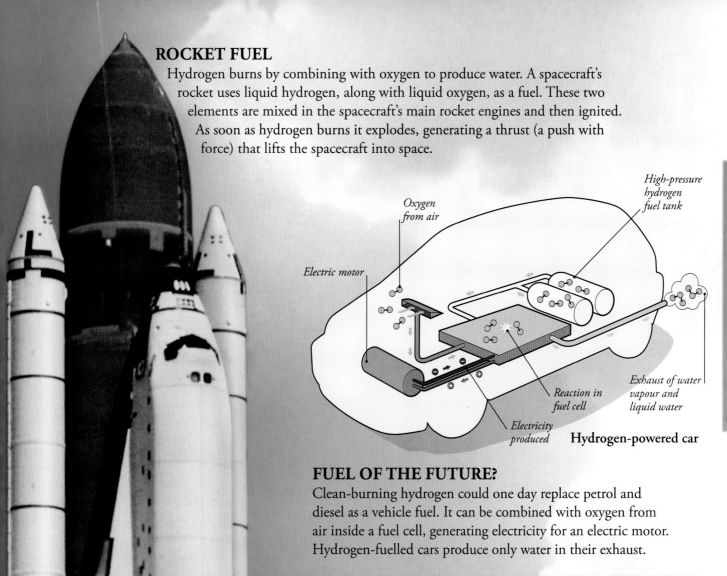

## ROCKET FUEL

Hydrogen burns by combining with oxygen to produce water. A spacecraft's rocket uses liquid hydrogen, along with liquid oxygen, as a fuel. These two elements are mixed in the spacecraft's main rocket engines and then ignited. As soon as hydrogen burns it explodes, generating a thrust (a push with force) that lifts the spacecraft into space.

*Oxygen from air*

*High-pressure hydrogen fuel tank*

*Electric motor*

*Reaction in fuel cell*

*Exhaust of water vapour and liquid water*

*Electricity produced*

**Hydrogen-powered car**

## FUEL OF THE FUTURE?

Clean-burning hydrogen could one day replace petrol and diesel as a vehicle fuel. It can be combined with oxygen from air inside a fuel cell, generating electricity for an electric motor. Hydrogen-fuelled cars produce only water in their exhaust.

### HYDROGEN DISASTER

In the 1920s–30s, gigantic hydrogen-filled airships carried passengers across continents and oceans. Then in 1937, the hydrogen in the German airship *Hindenburg* caught fire, probably from a stray spark. The massive fireball and explosion killed 36 people. Since then, airships have mostly used other less flammable gases.

**The *Hindenburg* explosion, 1937**

▲ US SPACE SHUTTLE *At lift-off, the enormous brown fuel tank of NASA's Space Shuttle contained 630 tonnes (705 tons) of liquid oxygen (LOX) and 106 tonnes (118 tons) of liquid hydrogen (LH$_2$), which reacted together to push the spacecraft away from the Earth.*

# Oxygen

Oxygen is vital to much of the life on Earth. Unless living things breathe it in every few seconds from the air around them, they cannot survive. Oxygen is also in every molecule of another vital substance – water – which all animals, plants, and other life forms need to stay alive. Oxygen from the atmosphere enables us to burn fuels such as wood and petrol, providing most of the energy for our homes and vehicles.

## CRITICAL FOR LIFE

All living things, including humans, use oxygen to turn food into energy that powers our bodies. Smaller land animals, such as insects, absorb oxygen from the air through their skins, while larger ones, such as humans, breathe it into their lungs. Water also has oxygen dissolved in it, which underwater animals absorb using gills.

### OXYGEN FACTS

- **Formula** $O_2$
- **Group** Non-metals
- **Density** 1.4 gm/l (0.18 oz/gallon)
- **Melting point** −219°C (−362°F)
- **Boiling point** −183°C (−297°F)
- **Main sources** Air, by cooling it into a liquid and then warming gently until oxygen gas is given off
- **Main uses** Needed by almost all living things; burning fuels in vehicles, heating, cooking, and industry; medical oxygen masks

Oxygen molecule

*Feathery red gills*

◄ GILLS OUTSIDE *Fish have gills under covers on the neck. The axolotl, a salamander-like amphibian, has frilly gills on the outside of its head.*

## BREATHING UNDERWATER

Human beings drown because our lungs cannot extract oxygen from water the way a fish's gills can. However, we can stay underwater for long periods by taking oxygen gas with us. Oxygen can be squeezed, or compressed, into metal tanks, so we can breathe it through a tube and face mask. This equipment is known as scuba (self-contained underwater breathing apparatus).

► CARRYING OXYGEN *Some scuba tanks contain gases with more oxygen than in normal air, so the diver can stay underwater for longer.*

## WOW!

Solid oxygen is pale blue, but you have to cool it to –219°C (–362°F) before it freezes.

## OXYGEN SHIELD

Normal oxygen gas has two atoms, $O_2$. There is another form, ozone, with three atoms, $O_3$. Ozone collects in a layer high above Earth's surface, where it is continually made and destroyed. The ozone layer protects us from the Sun's harmful ultraviolet rays. Human-made chemicals called chlorofluorocarbons (CFCs) destroyed some of this ozone, but it is slowly mending now.

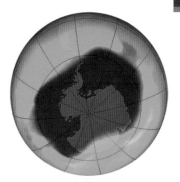

Ozone hole (dark purple) over Antarctica in 2006

## ESSENTIAL FOR BURNING

Any kind of burning or combustion needs oxygen. Molecules in the fuel break down as they join with the oxygen, producing light and heat. One way to stop burning is to prevent oxygen reaching the fuel. Spraying special foams containing non-oxygen gases on to the flames do this. So do fire blankets, gases such as carbon dioxide ($CO_2$), or other chemicals from a fire extinguisher.

## LIQUID OXYGEN

Oxygen becomes liquid below –183°C (–297°F), 10 times colder than a home freezer. Such low temperatures preserve medical samples for long periods. It is also a way to transport oxygen in a small space. Liquid oxygen (LOX) takes up 860 times less room than the gas, and it can be warmed later to turn it back into a gas.

◄ SUPER-COLD *Touching liquid oxygen to a living body part would make it freeze solid and snap off.*

▼ PUTTING OUT FLAMES
*Firefighters spray a foam blanket that stops oxygen in the air from reaching the fire.*

# Water

Water is the only substance on Earth that is commonly found in all three states of matter, as solid ice, liquid water, and gaseous water vapour. A water molecule is simple, with two atoms of hydrogen and one of oxygen. But it is perhaps the most important substance on Earth. Most of our planet's surface is covered with water and it makes up 60 per cent of our own bodies.

## WATER TO DRINK

Water without dirt or germs is essential for good health. To make it clean, dirty water is passed through a series of filters. These remove smaller and smaller items, from floating twigs and leaves, to bits of sand and grit, and finally some of the tiny germs. Adding chemicals such as chlorine then kills any remaining germs.

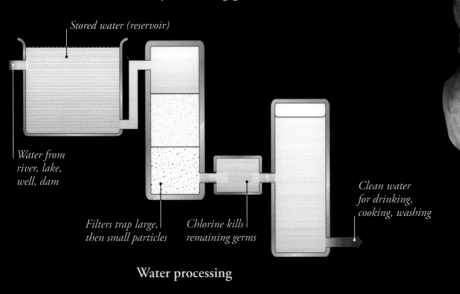

Stored water (reservoir)

Water from river, lake, well, dam

Filters trap large, then small particles

Chlorine kills remaining germs

Clean water for drinking, cooking, washing

**Water processing**

## WATER FACTS

- **Formula** $H_2O$
- **Group** Oxides
- **Density** 1 kg/l (8.34 lb/gallon)
- **Melting point** 0°C (32°F)
- **Boiling point** 100°C (212°F)
- **Main sources** Rivers, lakes, rain, melted snow and ice, purified from sea water by removing dissolved salts
- **Main uses** Needed by all living things to survive; for washing, cooking, and cleaning; for agriculture and farming; making electricity in hydroelectric dams; as a lubricant to reduce friction

**Water molecule**

## ICE FLOATS

Cooled down to 4°C (39.2°F), water gets smaller or contracts, and so it is heavier (more dense). Below 4°C (39.2°F) it starts to get bigger, or expands again. So when it freezes as ice at 0°C (32°F), it is lighter than the liquid water around it – and floats.

▶ HIDDEN DEPTHS *Ice just about floats, with up to 90 per cent of a giant iceberg below the surface.*

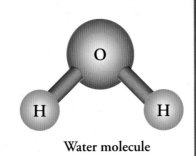

*Only a small part of the iceberg can be seen above the surface*

## DRYING UP

Water needs heat to "dry", that is, turn from liquid into gas. This is why puddles dry out faster on hot, sunny days. Our sweat is mostly water. This dries by drawing heat from the body, which keeps us cool in hot conditions.

▲ COOLING EFFECT *Hippos spend up to 16 hours a day submerged in rivers and lakes to keep their massive bodies cool.*

## CLOUDS AND WATER VAPOUR

The gas form of water, known as water vapour, is invisible. It can make up as much as 0.04 per cent of the air around us, but we cannot see it. What we can see is when this vapour cools and turns back into tiny droplets of liquid water, so small and light that they float – what we call clouds.

**Each cloud has trillions of water droplets.**

## PLANET WATER

Only one-third of planet Earth is covered with land, rocks, and soil. The rest is water. Almost 97.5 per cent of this is salty water, in seas and oceans. Less than 2.5 per cent is fresh water in rivers, lakes, soil, and rocks. And less than 0.01 per cent of this fresh water is easily available for us to use, in lakes and rivers.

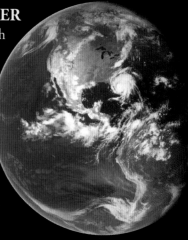

## WOW!

All the water on Earth gathered into a ball would measure 1,400 km (870 miles) across.

▲ OCEAN COVER *Almost half of Earth's surface is the water of the Pacific Ocean.*

47

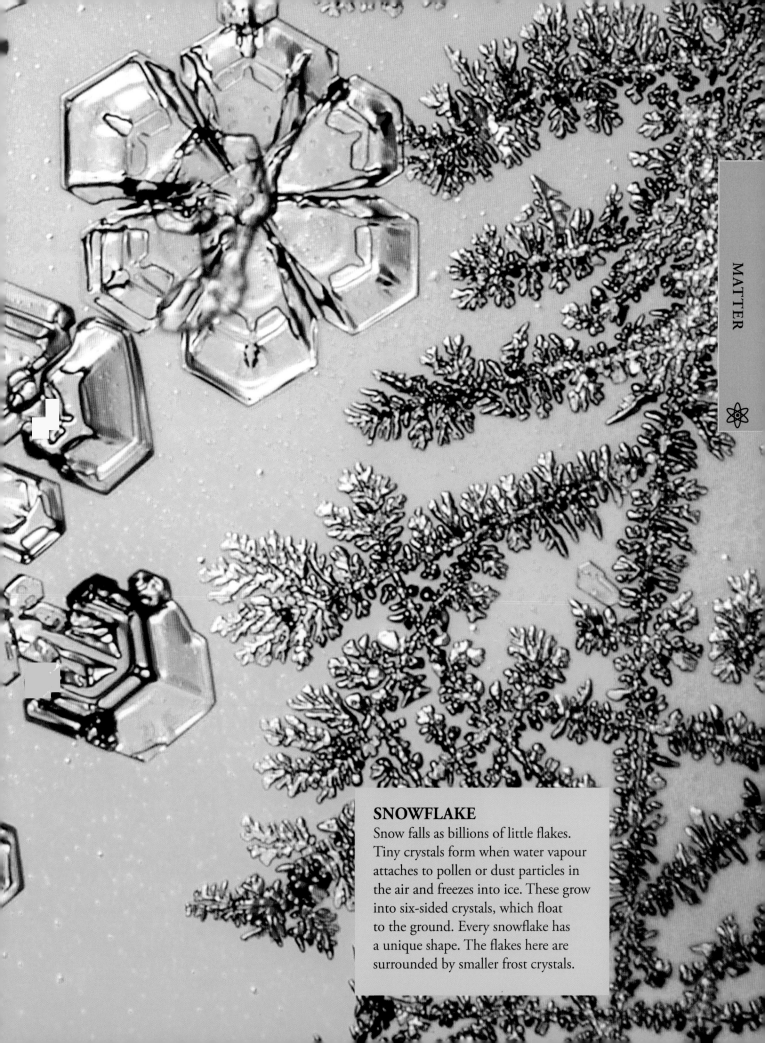

### SNOWFLAKE
Snow falls as billions of little flakes.
Tiny crystals form when water vapour
attaches to pollen or dust particles in
the air and freezes into ice. These grow
into six-sided crystals, which float
to the ground. Every snowflake has
a unique shape. The flakes here are
surrounded by smaller frost crystals.

# Nitrogen

Nitrogen is one of the most common chemical elements on Earth. It is all around us, making up 78 per cent of air. It is the fourth most common substance in the human body (after oxygen, carbon, and hydrogen), and it is found in similar amounts in all other living things. Nitrogen compounds are especially valuable in the soil, helping crops and plants to grow.

MATTER

## LIQUID NITROGEN

If nitrogen gas is cooled below −196°C (−321°F) it becomes liquid. In this form it is often used in extremely cold (or cryogenic) freezers, for storing blood, cells, eggs, seeds, and other parts of plants, animals, and humans. Liquid nitrogen is also used in science laboratories to cool super-fast electric motors, generators, and supercomputers.

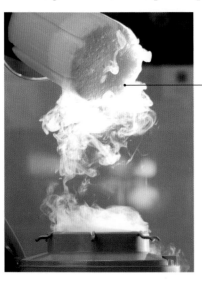

*These tissue samples have been frozen in liquid nitrogen.*

### NITROGEN FACTS

- **Formula** $N_2$
- **Group** Non-metals
- **Density** 1.25 gm/l (0.16 oz/gallon)
- **Boiling point** −196°C (−321°F)
- **Melting point** −210°C (−346°F)
- **Main sources** Atmosphere (air), minerals such as saltpetre (nitre), animal droppings, plant and animal bodies
- **Main uses** Food preserving; fertilizers and soil nutrients; making common industrial chemicals such as ammonia ($NH_3$) and nitric acid ($HNO_3$); explosives; super freezing

Nitrogen molecule

▼ BETTER CROPS
*Adding nitrogen-rich fertilizer to soil makes crops grow better.*

## IN THE SOIL

All plants need nitrogen compounds in the soil in order to grow. Certain plants "fix" nitrogen, which means they take in nitrogen gas from the air to make nitrogen-containing substances in their own bodies and in the soil. Farmers often add nitrogen-rich fertilizers to the soil to help crops grow.

## EXPLOSIVES

Many common explosives, including gunpowder, dynamite, and TNT, contain nitrogen compounds. Human beings have used these chemicals for many centuries to dig out tunnels and pits, and to open up mines and quarries to extract valuable minerals. They can also be used in explosive weapons, and in fireworks.

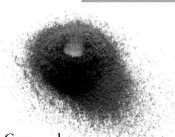

Gunpowder
(Potassium nitrate, $KNO_3$)

Dynamite
(Nitroglycerin, $C_3H_5N_3O_9$)

▲ HIGH-SPEED LANDING *Tyres filled with nitrogen stay pressurized for longer than air-filled tyres. They also cope well with changing air pressure as the aircraft climbs and descends to land.*

## KEEPING SAFE

Tyres in aircraft and some vehicles are filled with almost pure nitrogen, rather than air. Nitrogen does not leak through the tyre's rubber as easily as air does. Nitrogen also changes less in volume when the pressure outside changes as the plane rises and falls, and it contains less water vapour, which might damage the rubber.

## FUN WITH NITROGEN

Paintball guns and similar gas-powered devices use a quick blast of compressed (highly squeezed) gas from a tank to fire each paintball. At one time high-pressure carbon dioxide was a popular gas for this. Then nitrogen took over as it is safer to use. More recently compressed air, which is 78 per cent nitrogen, has become more common.

*Paintball fired by blast of high-pressure gas*

# Air

All around us is an invisible mixture of gases we call air. It contains about 15 different gases, although most of it is just two – nitrogen and oxygen. The exact contents of air vary from place to place. For example, damp air has more water vapour than dry air. Hazy air contains more floating dust particles than clean air. The blanket of air all around Earth is known as the atmosphere and this is where weather happens.

MATTER

## ENERGY FROM AIR

As the Earth and its atmosphere spin around once each day, the Sun warms different areas by different amounts. Warmed air is lighter and rises, so cooler air flows along to take its place. This moving air is wind – and it is a form of energy. We catch and use wind energy in many ways, from ancient windmills that grind grain and lift water, to the latest wind turbines producing electricity.

*Electricity generator in a pod*

*Rotor blade shaped to move around in wind*

*Blades held up on tall pylon*

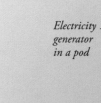

▶ WIND GENERATOR
*These turbines turn energy from the wind into electricity.*

▲ WIND PUMP *In dry regions, windmills work pumps to lift water from deep underground.*

▲ WINDSURFER *Wind pushes the sails of windsurfers, kitesurfers, yachts, and other craft.*

## WHAT IS IN AIR?

Almost all of air – about 99 per cent – is nitrogen and oxygen. The rest is made up of the noble gases (p.41), especially argon, but also neon, krypton, and xenon. The gas that varies most is water vapour, from almost 0 in desert areas to about 4 per cent where it is very moist or damp. Carbon dioxide and methane also vary with human activities such as burning fuels.

- Nitrogen 78%
- Oxygen 20.9%
- Argon 0.90%
- Other gases 0.17%
- Carbon dioxide 0.03%

## LESS AIR

The higher you go, the thinner the Earth's atmosphere becomes. By 5.6 km (3.4 miles) high the air pressure falls by half, then this halves again after another 5.6 km (3.4 miles), and so on. The temperature of the air also reduces with height, falling to −60°C (−76°F) by 10 km (6 miles), although it rises again to −5°C (23°F) by 50 km (31 miles). The atmosphere almost completely disappears by 100 km (62 miles) high – which is the official start of outer space.

▶ STAYING ALIVE *The air on top of the highest mountains is too thin to breathe, so climbers must carry oxygen in tanks.*

*Rotor angle changes with wind speed*

## NOISY AIR

The sounds we hear are made by vibrations in the air. These travel into our ears, where nerves pick up the vibrations and send signals to our brains. Our voices, musical instruments, and horns and sirens all make noise by causing vibrations in the air.

◀ LIGHT AND SOUND *As well as bright lights, lighthouses use loud fog horns to let people know where they are when fog makes it hard to see.*

## POLLUTED AIR

Smog is a mixture of smoke and fog. It forms in built-up areas where dust, smoke, fumes, and chemicals from cars and other vehicles, factories, and power stations all make the air hazy. Smog is worse when there is no wind, or when warmer air flows over the polluted air and traps it near the ground. It can cause breathing problems and other health hazards.

## WOW!

All the air in the atmosphere weighs 5.5 billion million tonnes (6.1 billion million tons), about one-millionth of Earth's weight.

▶ HAZY DAY *Bright sunlight worsens smog by making polluting chemicals in the air react together to produce dust-like particles.*

# Carbon

Carbon is the fourth most common element in the Universe. Unlike most other elements, it occurs in nature in several very different forms. By reacting with oxygen and other elements, it forms substances essential for life on Earth. It is also vital in our modern world for making all kinds of materials – from the hardest steels to the lightest fibres – and provides many of our main sources of fuel for cars, heating, and electricity.

## CARBON FACTS

- **Formula** C
- **Group** Non-metals
- **Density** About 2 kg/l (16.7 lb/gallon) – as solid powder
- **Melting and boiling point** Pure carbon is never found as a liquid – it turns from solid into gas at temperatures of 3,642 °C (6,588 °F)
- **Main sources** Mostly found in natural rocks, such as coal and diamond, and deposits of oil and natural gas
- **Main uses** Carbon compounds are used for fuel and making plastics

C

Carbon atom

## CARBON DATING

Some carbon atoms (known as carbon-14) are radioactive. Over thousands of years, these atoms break down, leaving ordinary carbon (carbon-12). By measuring the amounts of carbon-14 and carbon-12 in an ancient object, scientists can work out how old it is by a process known as carbon dating.

◄ HOW OLD?
*This bone came from a human being thought to have died in the Middle Ages. By taking a small sample, scientists can work out exactly how long ago the person died.*

## FOSSIL FUELS

All living things on Earth use carbon to build up their bodies. Coal, oil, and gas are the remains of the bodies of animals and plants that have been buried under the Earth for millions of years. We can extract these carbon-rich substances (called fossil fuels) and burn them to release energy.

► COAL MINING
*Giant excavators carve coal out of the ground. Coal is an important fuel, although burning it creates air pollution.*

## FORMS OF CARBON

Each carbon atom has four links, or bonds, which can be at slightly different angles to each other. This allows carbon atoms to join to each other in different ways to make very different substances. Coal, diamond, and graphite are all pure carbon, but they look and behave very differently because of the way the atoms inside them join together. Most forms of carbon are black and fairly soft, but diamond is clear, and is the hardest known natural substance.

In coal and soot, carbon atoms are all jumbled up.

Atoms in hard diamond are fixed in rigid boxes.

Soft graphite has atoms arranged in layers.

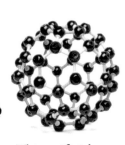

This artificial form of carbon is called "buckyball".

▲ STRONG, BUT LIGHT *Carbon fibre composites are used to make the bodies of racing cars and other speedy machines, which need to stay strong without weighing the vehicle down.*

## CARBON FIBRE

Carbon atoms can be joined into long chains of fibres that are 15 times thinner than a human hair, yet extremely strong. These carbon fibres are then added to other substances, such as plastics, to make very strong, yet lightweight materials known as composites.

WOW!

More than 8 billion tonnes (that's 8,000,000,000) of coal are mined around the world every year.

# Organic chemistry

The chemical element carbon is so important in living things that it has its own area of science, called organic chemistry. Carbon atoms can join with up to four other atoms, meaning they can be part of long and complicated molecules with very special properties. These are known as organic compounds. They are the building blocks of living plants and animals, fossil fuels such as oil and coal, and useful substances such as plastics, petrol, and medicines.

MATTER

## WOW!

There are 10 times as many organic substances as there are inorganic substances (those without carbon in them).

## THE CARBON CYCLE

Living things are about 20 per cent carbon. But there is only a limited amount of carbon in the world, so for new animals and plants to grow, carbon atoms have to be recycled. Dead plants and animals rot away in the soil, giving off carbon dioxide gas into the atmosphere. Plants absorb this carbon dioxide by a process called photosynthesis and turn it into food. Animals eat the plants, using the carbon to grow and give them energy. They release carbon dioxide when they breathe out. Carbon is also trapped in oceans and in fossil fuels underground.

**Sunlight**

*Plants absorb more carbon dioxide than they release*

*Animals breathe out carbon dioxide*

**Plants and trees**

*Animals take in carbon by eating plants*

**Animals**

*Dead animals release carbon dioxide as they decompose*

*Dead plants may decompose or be preserved as fossil fuels*

**Rocks**

▶ ROUND AND ROUND *Up to 100 billion tonnes of carbon move through the carbon cycle every year between plants, animals, soil, and water.*

56

## CARBON CHAINS

With four chemical bonds each, carbon atoms can form long chains joined to each other, and to other atoms. Many of these long molecules have the same small group of atoms repeated hundreds, even millions, of times. These giant molecules are known as polymers, and are used in plastics.

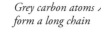

*Grey carbon atoms form a long chain*

▶ CARBON STRINGS *The human-made plastic nylon is a carbon polymer made from mixing organic liquids together.*

**Carbon dioxide in the atmosphere**

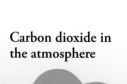

*Oceans absorb carbon dioxide, but sea life also breathes it out*

*Burning fossil fuels releases carbon dioxide*

*Skunks use smelly organic molecules to drive off attackers*

Oceans

Factories

*Humans dig up fossil fuels and burn them for energy*

*Carbon from ancient animals and plants forms fossil fuels*

## SMELLY CHEMICALS

Many organic substances have strong smells, such as sweet-scented flowers, strong herbs and spices, and the stench of a skunk's spray. This is partly because the compounds are volatile – that is, they turn into a vapour at normal temperature. They float through the air and into our noses.

## CARBOHYDRATES

Carbon molecules in plants can join with oxygen and hydrogen to make carbohydrates, such as sugar and starch. Many animals, including humans, rely on carbohydrates in their diet, as they are easily broken down in the body to give us energy.

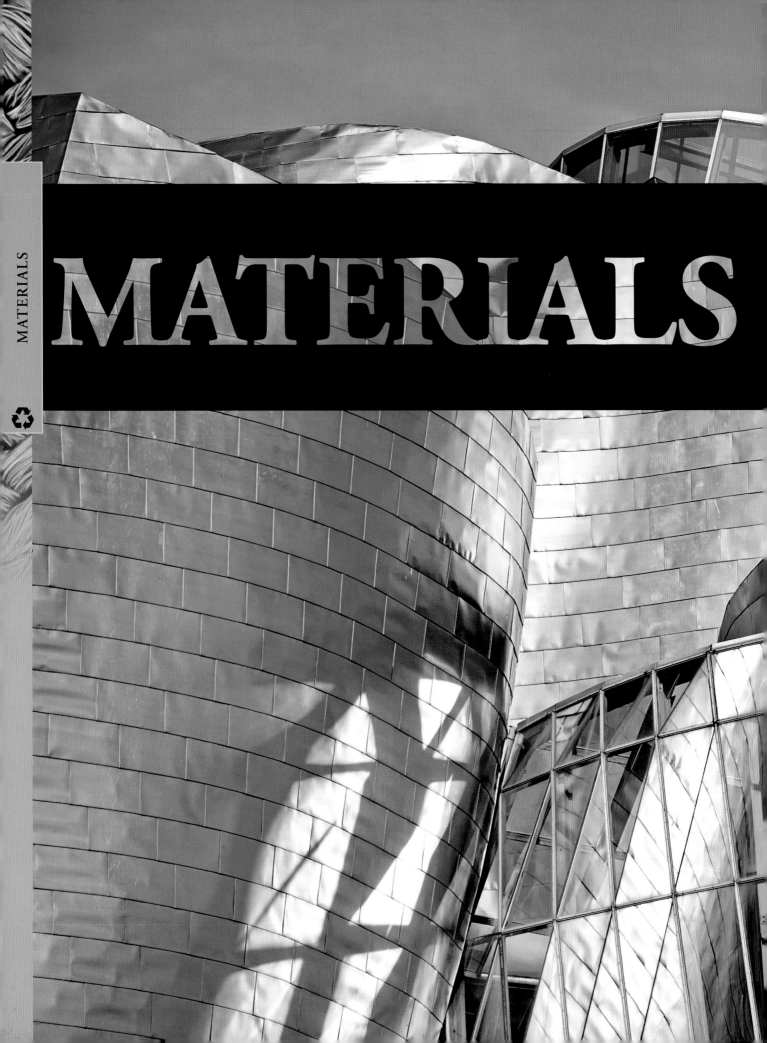

# MATERIALS

Materials are substances we can use to make things. From hard stones to magnetic metals, to soft plastics and synthetic fabrics, different materials are useful for different tasks.

**BUILDING MATERIALS**
Architects use different kinds of material to make their buildings look impressive and stay strong. The Guggenheim Museum in Bilbao, Spain, is decorated with interconnected panels of stone, glass, and titanium metal.

# Defining materials

Materials are substances we use to make things. They range from hard metals to soft fabrics, natural wood to human-made plastics. Different materials have different properties, which make them useful in different ways. Engineers and inventors are experts at choosing the right material for every job.

## INNOVATIVE USES

Most buildings consist of several different materials, such as brick, glass, concrete, and wood, each chosen for its features and qualities. However, traditional materials can also be adapted for new uses. For example, cardboard may be treated to make it waterproof and flame-resistant, and then used as a construction material.

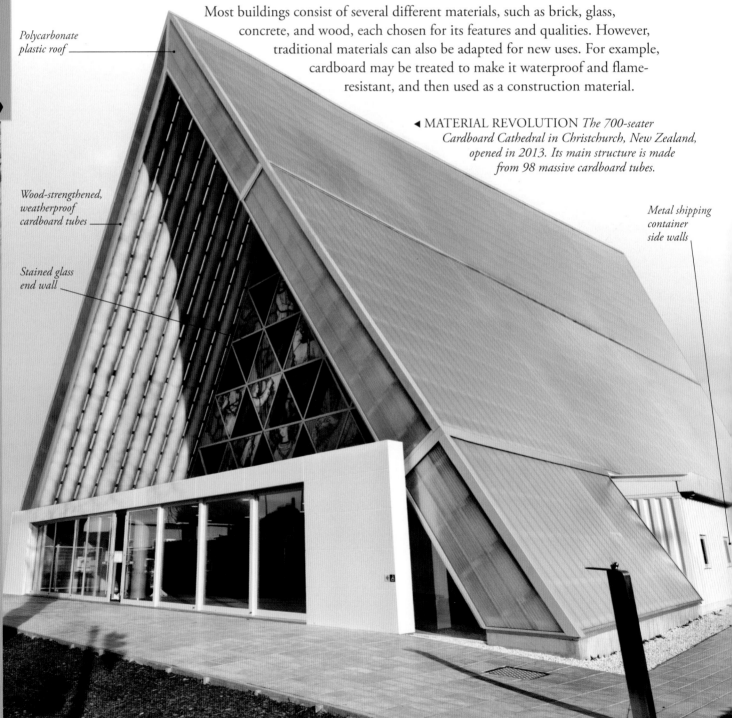

◄ MATERIAL REVOLUTION *The 700-seater Cardboard Cathedral in Christchurch, New Zealand, opened in 2013. Its main structure is made from 98 massive cardboard tubes.*

Polycarbonate plastic roof

Wood-strengthened, weatherproof cardboard tubes

Stained glass end wall

Metal shipping container side walls

## CHOSEN FOR LOOKS

The appearance of a material can be as important as its strength or hardness. For example, stone blocks for buildings are carefully chosen for their colours, patterns, and how well they polish into a shine, as well as for their long-lasting strength. Different kinds of woods also have their own individual appearance, from white birch to black ebony.

▲ MARBLE WONDER *The Parthenon, an Ancient Greek temple in Athens, Greece, is more than 2,400 years old. Its tall columns are made from a gleaming white rock called marble.*

*Flint was shaped and used as a cutting tool*

## FIRST MATERIALS

The first materials ever used by people were natural stone, wood, and animal parts such as bones and horns. The earliest stone tools date back more than 2 million years. Wood is easier than rock to shape, carve, and paint, although it does not last as long. However, some very old wooden spears and tools, dating back more than 200,000 years, have been found preserved in bogs.

*Wooden handle*

◄ EARLIEST TOOLS *In the prehistoric times, stones such as flint, basalt, and sandstone were used to make hammerstones, hand axes, and other large cutting tools.*

*Concrete slab base*

## WOW!

The material called Wurtzite Boron Nitride (WBN) is even harder than diamond, but you need an explosion to make it!

## PROPERTIES OF MATERIALS

Each material has some features, called properties, that make it suitable for certain uses. For example, is it hard or soft, strong or weak, stiff or bendy? The ability to be shaped by hammering, as with red-hot metal, is called malleability. A material that carries heat well is a conductor, while one that does not is an insulator.

| PROPERTY | EXAMPLE |
|---|---|
| Chemically reactive | Sodium |
| Chemically unreactive | Glass |
| Electrical conductor | Copper |
| Electrical insulator | Ceramic |
| Magnetic | Iron |
| Non-magnetic | Wood |
| Hard | Diamond |
| Soft | Clay |
| Dense | Granite |
| Light | Foamed plastic |
| Flammable | Paper |
| Non-flammable | Stone |
| Rigid | Steel |
| Flexible | Rubber |

# Plastics

The first synthetic (entirely human-made) plastic was produced in the early 1900s. Today it is one of our most useful and adaptable materials. There are hundreds of different kinds of plastics, in all sorts of different colours, and with all sorts of different properties, from strong and hard to soft and flexible. Most plastics melt when they are heated, so they are easy to mould into any shape. This makes them useful for many different things.

## SOURCE OF PLASTIC

Most modern plastics are made from crude oil, which is extracted from deep underground. The oil is separated into different chemicals, which can be turned into plastics. When the world runs out of crude oil, we will have to rely on plastics made from plant material (bioplastics).

▲ BASIC BEADS *Plastics are often produced as pellets or beads, which can be melted down to make useful products.*

## PLASTICS EVERYWHERE

Plastics are extremely useful, and our lives would be very different without them. Plastics are waterproof and do not rot away, making them great for storing liquids. They can be easily cleaned, making them safe and hygienic for preparing and storing food. They can be made into almost any shape, so can be used to make machines and toys. And they can take on bright colours in almost any shade.

## SOFT AND HARD PLASTICS

Plastics usually contain several other chemicals, called additives, to give them different properties. Some are mixed with hardeners so they are tough and resist scratching. Others have softener additives so they will squash and bend easily, then spring back to their original shape, like natural rubber.

Soft plastic toy

Industrial hard plastic pipes

## SEE-THROUGH PLASTIC

Lighter than glass, and much less fragile, transparent plastics are ideal for everything from windows to water bottles. We can see through these plastics because of the way their molecules are arranged. All plastics are made up from long chains called polymers. If these polymers all line up side-by-side, the plastic becomes see-through.

### PROPERTIES

- Adaptable – can take on many different shapes, colours, and properties
- Easy to shape – can be melted and re-shaped again and again
- Flexible – able to bend or squash rather than splinter or snap
- Waterproof – good for storing water or keeping things dry
- Insulating – protect against electricity, and keep things warm or cool

## PLASTIC RECYCLING

Most plastics are long-lasting and do not rot away. This makes them very useful, but also causes problems when we no longer need them. Unlike wood and natural textiles, plastics stay around when we throw them away, building up in rubbish heaps. The best way to deal with this is to melt down used plastic and turn it into new products we can use again. This process is called recycling.

▶ PLASTIC FOR RECYCLING *Instead of throwing away old plastic bottles, we can melt them down to make new products.*

MATERIALS

# Glass

People have been making glass for thousands of years by heating up special kinds of sand mixed with other chemicals. For centuries, glass was the only see-through material available, so it was used in the earliest windows and lenses. It is also easy to shape when it is heated up, and does not rot or melt away in water, making it useful for bottles and vases. But glass cannot be used everywhere because it is very fragile, and can be dangerous if it breaks.

## BEAUTIFUL OBJECT

Glass can be blown into almost any shape, and pigments (coloured dyes) can easily be added to give it special colours. Humans have been using glass to create beautiful ornaments and decorations ever since it was first invented. Clever glass-makers can even create shapes inside other shapes, as in this vase.

## MAKING GLASS

When glass gets hot enough, it becomes a thick, sticky liquid which can be moulded into different shapes. Today, most glass objects are made in factories. In the past, glass workers used to blow down long metal tubes to create bubbles of glass which they could shape into bottles and vases.

◄ GLASS BLOWING
*These glass workers are shaping globs of molten glass by blowing air into them. The glass has to be kept at exactly the right temperature.*

## SAFETY GLASS

Car windscreens need to be hard and see-through, without breaking into dangerous pieces if there is a crash. Most windscreens are made of "laminated" glass, which is two sheets of glass with a thin layer of clear plastic sandwiched between them. If the glass breaks, the pieces stay stuck to the plastic, so there is much less danger from flying shards. As an extra safety measure, modern glass is often specially treated so that it breaks into small, square lumps rather than long, sharp shards.

### PROPERTIES

- Long-lasting – resists water, heat, cold, damp, decay, and rot
- Transparent – but can also be made opaque to let in light without letting you see through to the other side
- Easy to shape – can make everything from giant windows to tiny lenses
- Unreactive – not affected by most chemicals, even strong acids
- Does not conduct electricity

## REFLECTIVE GLASS

Certain kinds of chemical coating can turn glass into a mirror, reflecting light coming towards it. The other side of the glass can also be a mirror – or just ordinary glass. This can be used to make windows that are easy to see out of but hard to see in through.

▲ MIRROR WINDOWS *Many modern buildings have whole walls of glass windows hung on a metal frame, making the insides feel light and airy.*

▲ ANCIENT GLASS *Early glass beads, bowls, and vases were made in bright colours, but it was hard to make glass pure and polished enough to be completely see-through.*

## LONG-LASTING

Glass is very hard and very resistant to water and other chemicals, so it can last for thousands of years. The oldest glass objects include beads dating back more than 4,000 years. Many early glass objects were made for their beautiful colours, and were used in jewellery. Glass windows appeared around 2,000 years ago, but only in the most important buildings.

# Ceramics

Some special kinds of clay will set as hard as rock if they are baked at very high temperatures. These are known as ceramics, and people have used them for more than 25,000 years to create pots, statues, and tools. The baking process is called "firing", and takes place in a special oven called a kiln. Today, ceramics are used everywhere from china plates to spaceship shields.

*Ceramic insulating sleeve*

**Spark plug**

### CERAMICS IN MACHINES
Ceramics can be moulded into very precise shapes when soft, and are very hard and tough when fired. They do not conduct electricity, so are used in power lines and engine spark plugs to make sure electrical current does not flow to the wrong place. However, ceramics are brittle, so cannot be used in machines where they would crack or splinter.

### CERAMICS IN ART
Ceramics have been used in art for many thousands of years. Early humans discovered that pigments (coloured dyes) could be painted on the clay before it was fired. The heat of the kiln bakes the colours on to the clay, so the pattern never washes away.

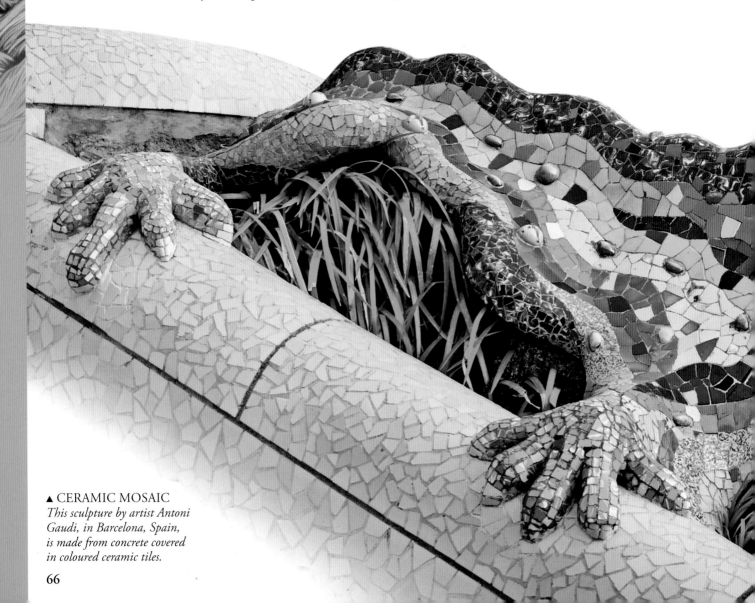

▲ CERAMIC MOSAIC
*This sculpture by artist Antoni Gaudi, in Barcelona, Spain, is made from concrete covered in coloured ceramic tiles.*

MATERIALS

## CERAMICS FOR PROTECTION

Ceramics fired at very high temperatures, up to 1,200°C (2,192°F), can resist enormous temperatures afterwards. They are used to line ovens, furnaces, stoves, and kilns. They are also used in spaceships, which have to withstand enormous heat as they re-enter Earth's atmosphere. A layer of ceramic tiles and other heat-proof materials is used to cover the bottom of the craft so that it does not burn up on its way home.

▶ HEAT SHIELD *The underside of this space shuttle is covered with a shield of more than 24,000 black, heat-resistant ceramic tiles.*

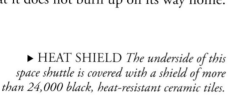

### PROPERTIES

- Long-lasting – resist rain, heat, cold, damp, decay, and rot
- Unreactive – not affected by most chemicals, even strong acids
- Insulating – do not allow electricity to pass through
- Stiff – do not bend under strain (but may crack and splinter)
- Easy to shape – and easy to colour for different uses

## CERAMICS AT HOME

Ceramics are smooth, shiny, and easy to clean. They do not burn and they do not react with food and drink. All this makes them ideal for plates, bowls, mugs, and all kinds of cooking and eating utensils. They are also commonly used for bathroom fittings, such as toilets, sinks, and baths, because they can be washed with harsh, germ-killing chemicals without being damaged.

## CERAMICS IN THE BODY

Ceramic materials are used to make some artificial body parts, such as teeth and joints. Since ceramics are hard but brittle, there is always the danger they might split, so false teeth are usually made with additives to help them absorb pressure and strain without cracking.

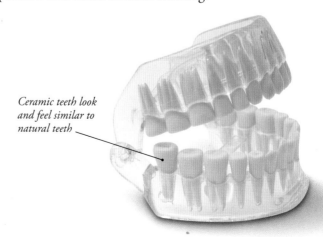

*Ceramic teeth look and feel similar to natural teeth*

# Synthetic fibres

For thousands of years, people have made fabrics and textiles from natural fibres such as cotton from plants, silk from silkworms, and wool from sheep. In the 1890s, two new fabrics, viscose and rayon, were produced by breaking down fibres in wood. The first fully synthetic fibre, entirely made by humans, was nylon in 1935. There are now dozens of synthetic fibres, each with its own combination of strength, thickness, and fluffiness.

## MAKING FIBRES

The first stage is usually to mix various chemicals and squeeze the resulting liquid through tiny holes or spinnerets. Out come long, thin strings that turn solid and become bendy fibres. These are twisted and wound together to make ropes, yarn, and threads, which are woven into textiles, knitted, knotted, and combined in other ways.

◄ FIRST-STAGE FIBRES *Bundles of single fibres are drawn together into long, loose strands before further treatment.*

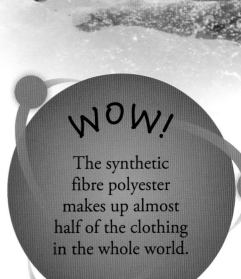

WOW!

The synthetic fibre polyester makes up almost half of the clothing in the whole world.

## MICROFIBRES

Synthetic fibres can be made in various thicknesses, from several millimetres to thousands of times thinner. Microfibres are 100 times narrower than human hairs. Fabrics made from them are usually soft and bendy. Some are made to soak up water and other fluids, so they are suited to cleaning and polishing. Others resist water, as in microfibre jackets and similar garments, and types of fleeces.

▲ WOVEN TOGETHER *A powerful microscope shows how bundles of microfibres are woven over-and-under into a strong, thick fabric.*

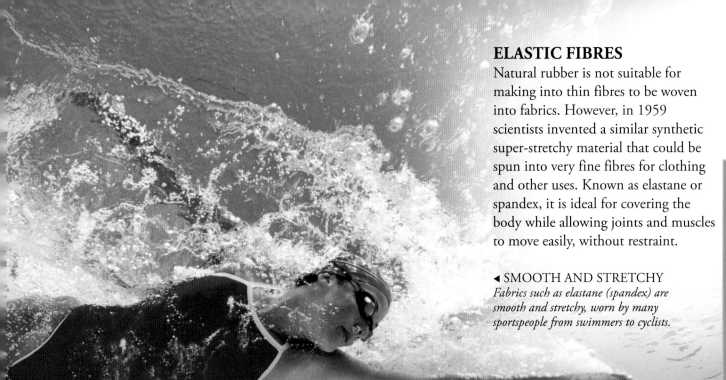

## ELASTIC FIBRES

Natural rubber is not suitable for making into thin fibres to be woven into fabrics. However, in 1959 scientists invented a similar synthetic super-stretchy material that could be spun into very fine fibres for clothing and other uses. Known as elastane or spandex, it is ideal for covering the body while allowing joints and muscles to move easily, without restraint.

◄ SMOOTH AND STRETCHY
*Fabrics such as elastane (spandex) are smooth and stretchy, worn by many sportspeople from swimmers to cyclists.*

## LIGHT AND WATERPROOF

Most modern tents are made from synthetic fabrics, for example nylon and polyester, often coated with other chemicals such as silicone or polyurethane. They do not absorb water or rot, are lightweight, and keep out rain.

**Nylon tent**

## PRODUCTS AND USES

There are dozens of different kinds of synthetic fibres, some more commonly used than others. The compounds used to make synthetic fibres come from raw materials such as petrochemicals. Each kind has many uses and has one or more common names, such as spandex or elastane. These are produced to slightly different qualities by different makers, and combined with other chemicals or fibres to give a variety of trade names, for instance Lycra® and Creora® for elastane.

◄ ACRYLIC *Strong, lightweight, and warm with a wool-like feel, acrylic is used for warm outer garments, as well as furnishings and carpets.*

◄ RAYON *Known as "artificial silk", rayon fibres are smooth and soft. They absorb heat and repel water, making them ideal for umbrellas.*

◄ POLYESTER *Tough and durable, this fabric has many daily uses, such as clothes and blankets and industrial uses such as conveyor and safety belts.*

◄ NYLON *The world's first synthetic fibre, nylon is a strong fibre, which has widespread uses from parachutes and tents, to bags and ropes.*

# Composites

A composite material is made from two or more substances combined, which are usually quite different from each other. The aim is to have the best features of each substance, such as hardness and strength, so that the composite is better overall than its ingredients. Scientists invent new composites almost every week for specialized uses. However, some composites, such as papier-mâché and concrete, are thousands of years old.

## BEST COMBINATION

Concrete is very strong, but crumbles easily if it bends. To prevent this, steel bars are often set into concrete in buildings, making a strong composite that does not crumble.

*Metal bars are flexible*

*Concrete is stiff and makes firm base*

## FIBREGLASS

Fibreglass is a composite material made of soft, flexible plastic with very thin strands of glass inside. It is also known as glass-reinforced plastic, or GRP. The glass helps the plastic hold its shape, while the plastic stops the glass from breaking.

▼ SPEEDBOATS *Many speedboats are made of glass-reinforced plastic (GRP). It is waterproof, can be shaped and smoothed, yet can also resist knocks, bends, and twists.*

## CONCRETE

One of the oldest human-made composites, concrete is a mix of cement, sand, small stones, and water. It sets like rock, thanks to a chemical reaction (see p.32). Stones give it a strong overall framework, sand fills the small gaps between them, and cement bonds or glues them all together.

▲ ROMAN CONCRETE *The Colosseum in Rome, Italy, was an early concrete structure built almost 2,000 years ago.*

## SUPER-STRONG FABRIC

Some synthetic fibres (see p.68) are immensely strong and tough. Weaving them with other, more flexible fibres makes a composite that is strong but soft. With brand names such as Nomex® and Kevlar®, these fabrics are made into all kinds of super-tough materials, from bullet-proof clothing to sword-fencing outfits, ropes, tyres, and sails.

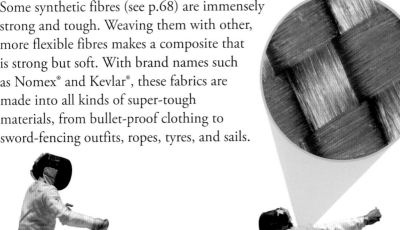

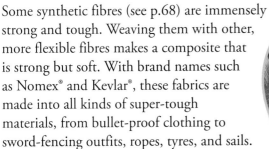

WOW!

Kevlar® is five times lighter than steel, but can survive five times as much force.

◄ PROTECTIVE GEAR *The protective clothing worn in fencing is made of Kevlar®. It protects the fencers, and is light and flexible too.*

## HARDEST COMPOSITES

Some modern tanks and armoured vehicles are protected by composites. They often have hard ceramic plates surrounded by layers of metal and stretchy plastics. This "multi-sandwich" absorbs and spreads the impact of explosions and bullets, so that the people inside remain unharmed.

► POWER DEFENCE *The 54-tonne (60-ton) US M1 Abrams tank is protected by layers of composite armour.*

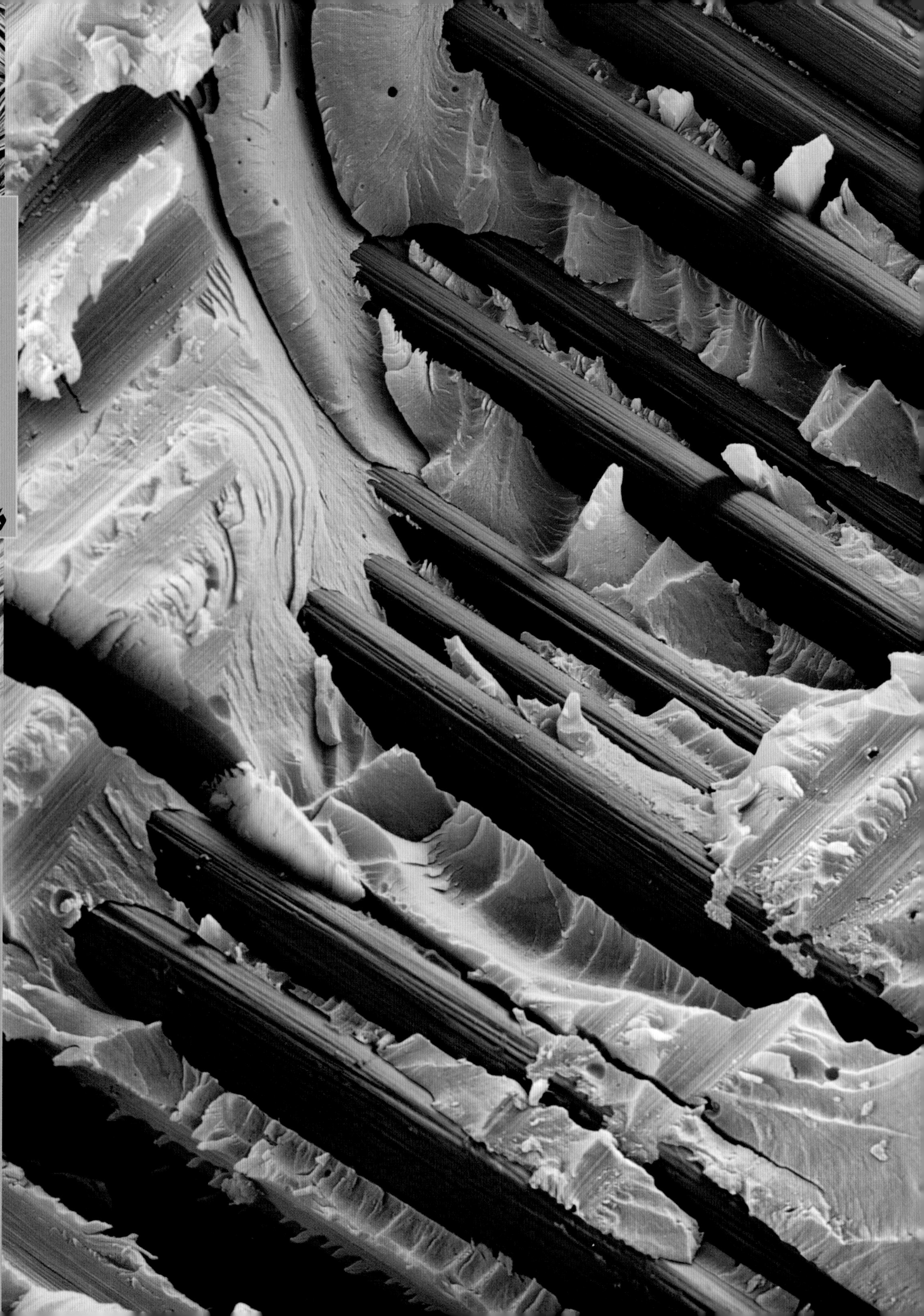

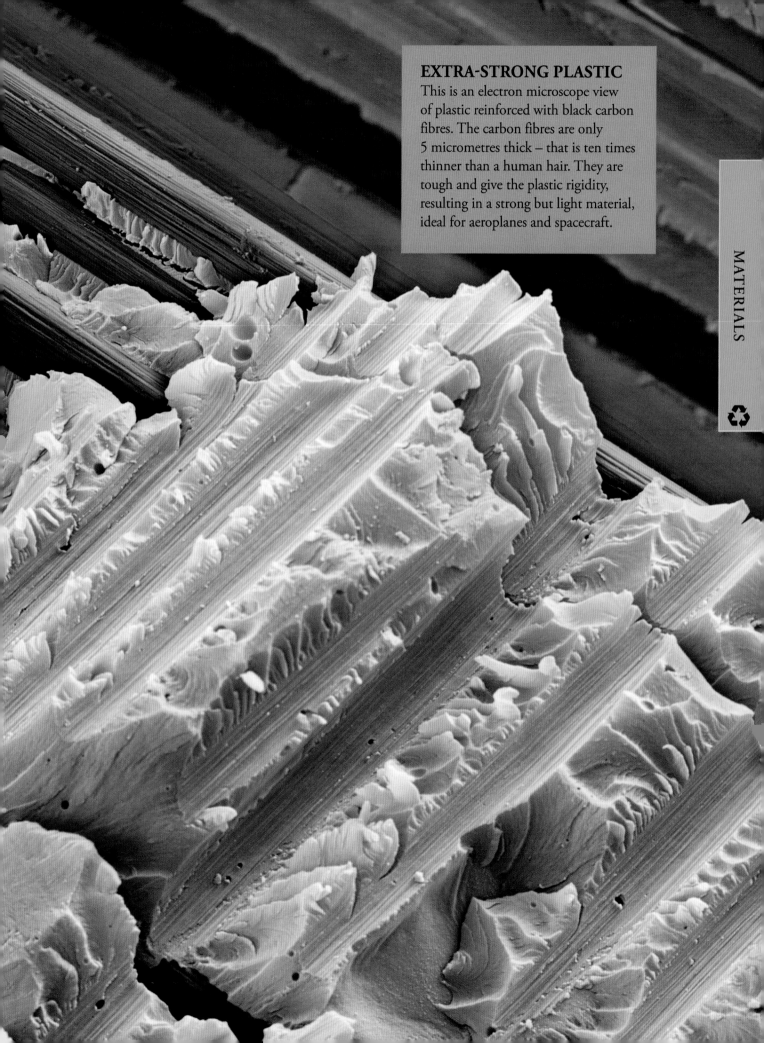

**EXTRA-STRONG PLASTIC**
This is an electron microscope view of plastic reinforced with black carbon fibres. The carbon fibres are only 5 micrometres thick – that is ten times thinner than a human hair. They are tough and give the plastic rigidity, resulting in a strong but light material, ideal for aeroplanes and spacecraft.

# Earth's resources

All the materials we use, and the things we make from them, come from the Earth. Some are natural and are used in their raw form, such as wood, stone, and ceramics. Others, such as synthetic fibres and human-made plastics, are made by processing natural resources like crude oil and minerals. The energy to make all these products also mainly comes from the Earth, as fossil fuels. These resources cannot last for ever and we cannot put most of them back.

**WOW!**

Earth receives more energy from the Sun in one hour than everyone in the world uses in one year.

## FOSSIL FUELS

As the most easily-reached sources of coal, gas, and oil are being used up, miners and drill rigs must go to remote mountains, deep oceans, and icebound seas in search of more. At today's rate of use, known coal reserves may run out within 120 years, while oil and gas reserves will last less than 80 years.

▼ OIL PLATFORM
*Huge exploratory drill platforms search for oil under the sea bed.*

## TIMBER

About half of all forests are grown and harvested in a sustainable way for paper, timber, and fuel. But more than half of Earth's tropical rainforests have been cut down in the past 100 years, for rare timber and farmland. Almost none of these tropical forests have been replanted.

▲ CONTINUING USE *In sustainable forests the timber is removed without too much damage, then new trees are planted.*

## WATER

As Earth's population rises, humans need more clean water for drinking, cooking, and washing. We also use water for growing crops and raising farm animals. Water has to be cleaned and purified before we can use it, which requires energy and resources.

▲ SALTY TO FRESH *Salt can be removed from sea water to make it fresh and clean, but this uses vast amounts of fuel or energy.*

*Drill pipes inside derrick tower*

## FARMLAND

Farmers need fertile soil to grow crops and raise animals. Pollution or over-farming (growing too much of the same crop in the same place) can damage the soil, making fertile land a resource that needs to be protected.

▲ FARM TERRACES *Levelled strips along a slope, called terraces, help to hold water and prevent soil from being washed away.*

## ENERGY

Humans around the world need energy to power cities, vehicles, homes, and factories. Energy is another kind of resource. Today, most of it comes from burning fossil fuels, but in the future we will need to find cheaper, less polluting energy sources.

▲ WIND FARMS *The wind produces a supply of non-polluting energy. However, at the moment it can only supply a small amount of the energy we use.*

# Materials in industry

Millions of factories around the world take in resources such as energy and raw materials, and churn out chemicals and products, ranging from fertilizers to jumbo jets. One of the main resources is oil, which is used by 90 per cent of all the world's industries. Other resources include natural gas, seawater, and minerals.

## INSIDE THE FACTORY

Chemical factories, or plants, start with raw materials, which are mixtures of hundreds of different substances. The raw materials are separated using physical processes such as evaporation and chemical reactions. For instance, an oil refinery receives crude oil, a complex mixture, which it separates into more than 200 different substances. These can then be used for making various products.

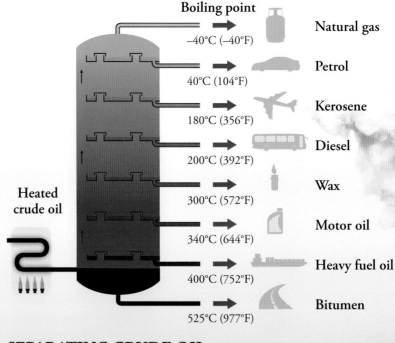

**Boiling point**

| | | |
|---|---|---|
| –40°C (–40°F) | → | Natural gas |
| 40°C (104°F) | → | Petrol |
| 180°C (356°F) | → | Kerosene |
| 200°C (392°F) | → | Diesel |
| 300°C (572°F) | → | Wax |
| 340°C (644°F) | → | Motor oil |
| 400°C (752°F) | → | Heavy fuel oil |
| 525°C (977°F) | → | Bitumen |

Heated crude oil

▶ METAL MAZE *Oil refineries are among the world's biggest industrial sites. Some are the size of towns and employ thousands of people.*

## SEPARATING CRUDE OIL

Crude oil is heated in a tall tower called a fractionation column. The chemicals inside turn to gas and rise up the column. As they get higher, they cool down, turning back into liquids at different heights, so they can be separated out.

▼ WARNING!
*Signs are used on lorries around the world to warn of dangerous materials, such as those which are poisonous or explosive.*

FLAMMABLE
GAS
2

## SUPPLY AND TRANSPORT

Specially designed vessels called tankers are used for transporting chemical products in bulk. Numerous safety rules regulate the transport of these products. Because of the danger of fire or explosion, these lorries carry markings to show what's inside.

## DANGERS AND DISASTERS

Chemical factories can be dangerous places if things go wrong. Chemicals can leak out, polluting nearby air, soil, and water. Radiation can escape from some facilities, while others might see fires or even explosions. That is why all factories are carefully controlled to minimize the chances of an accident.

▶ AFTERMATH
*In 2013 a fire and explosion took place at the West Fertilizer Factory, Texas, USA, damaging nearby homes.*

▶ PERSONAL PROTECTION
*Industrial workers wear protective clothing to keep them safe in case of accidents.*

## HEALTH AND SAFETY

Industrial materials and chemicals can be far stronger and much more harmful than those used in homes. Safe practices include storing them in tanks, barrels, or similar secure containers that won't leak, which are clearly marked and kept away from possible causes of heat and fire.

MATERIALS

# Recycling

Since Earth's raw materials, energy, and other resources will not last forever, it is essential to make sure they are not wasted. We can do this by repairing, reusing, and recycling. Recycling usually begins by sorting items into separate groups based on their main materials, such as metals, glass, plastics, and paper. Recycling saves not only Earth's resources, but also an immense amount of energy.

## RECYCLING CENTRE

To make recycling efficient, different materials must be separated from each other. You can do this when you throw things away, by putting materials to be recycled into separate bins. Plastics are often sorted by experts, since different plastics need different recycling methods, and it can be hard to tell them apart.

## LANDFILL

Rubbish tips or landfill sites bring problems such as smells, litter, pest animals, and possible pollution. They are costly to cover over and make safe. Some plastics can last forever under the ground without rotting away. Throwing away recyclable materials means they are wasted and cannot be reused.

▲ AROUND FOR CENTURIES *Even after landfills are covered with soil and plants, the rubbish may cause future problems.*

▼ PLASTIC PICKERS *Workers pick out and sort different kinds of plastics so they can be recycled the right way.*

## PLASTICS

Each type of plastic has its own features (see pp.62–63). However, its chemical make-up means it must be recycled in a particular way. For some plastics, recycling is not practical with today's technology, and burning them gives off harmful fumes. The solution is to use less plastic and reuse where possible.

1 PETE Polyethylene Terephthalate

2 HDPE High Density Polyethylene

3 V Vinyl

4 LDPE Low Density Polyethylene

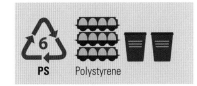
5 PP Polypropylene

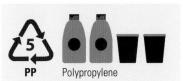
6 PS Polystyrene

7 OTHER Other

## GLASS

One of the best materials to reuse and recycle is glass (see pp.64–65). It is hard, smooth, and easily cleaned, so glass jars and bottles are ideal for reuse and filling with other contents. Recycling glass saves more than one-third of the energy and one-half of the raw supplies needed to make new glass, and it can be done many times over.

▲ BACK INTO PRODUCTION *The same glass can go round and round the recycling process dozens of times. In a landfill, it will last a million years.*

## RECYCLED PRODUCTS

Glass, metals, and many plastics can be melted down to make new objects that are just as good as non-recycled ones. Recycled paper is often less smooth than new paper, but is ideal for toilet paper and kitchen towels. Recycled fabrics can be used to make cleaning rags and dust sheets.

◄ RECYCLED CHAIR *Plastics can be melted down and formed into new shapes, from plastic bags to garden chairs.*

▲ PLASTICS BY NUMBERS *The numbers in the recycling triangle symbols are codes for the type of plastic used in the items.*

## REUSE AND UPCYCLE

Selling, swapping, or giving away saves even more materials, energy, and landfill space than industrial recycling. There are many ways to reuse such as swap shops, charity shops, car boot sales, and on websites. In "upcycling", unwanted items are reused to give extra value, for example, using bits of metal from old machinery to create works of art or craft-made utensils.

▶ SORTING FOR REUSE *Almost any useable item could be wanted by someone, somewhere. Every little bit helps to save Earth's resources, decrease waste, and reduce pollution.*

# Future materials

As scientists and engineers develop new techniques for combining atoms, more and more materials become available for us to use. Some of the latest designs have amazing properties, from metals that can float on water, to armour plating that only goes solid when an impact occurs. These new inventions could transform our lives in years to come.

## INTELLIGENT MATERIALS

Smart or intelligent materials can change their properties depending on conditions around them. You might have seen them in glasses that look clear in normal light but darken in bright sunlight. A special form of aluminium has been created that turns see-through when electrical current is passed across it. This could be used to make buildings with windows stronger than glass that can be made opaque at the flick of a switch. Other possibilities include clothing that changes with temperature to keep you warm or cool in any weather.

## ARMOURED FABRICS

Future protective gear may feel like ordinary clothes. It will be lightweight, flexible, and waterproof, making it comfortable to wear. However, when knocked or hit, it will instantly stiffen into a tough shell to protect the body beneath and spread the impact harmlessly over a wide area. After the impact, the fabric will return to normal.

▶ WELL PROTECTED *Flexible armour could allow motorcyclists to stay cool and comfortable while riding, at the same time keeping them safe.*

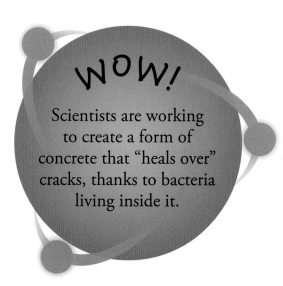

◀ SPACE BASE *A future base on the planet Mars could use smart materials to protect the people inside.*

**WOW!**
Scientists are working to create a form of concrete that "heals over" cracks, thanks to bacteria living inside it.

## USING ENERGY

Electricity is a very convenient and adaptable form of energy. It is easily transported long distances along wires, and it can be changed into many other energy forms for our use. At the moment, we rely on fossil fuels to generate most of our electricity, but these cause pollution and will run out in the future. One of the great challenges for future scientists will be to find materials that can store and transport electricity more easily, as well as to find new ways of capturing energy from renewable sources such as the Sun or the wind.

◀ BATTERY POWER
*As petrol becomes more expensive, we may rely on electricity to power our cars. New materials may help us to build better batteries, which will allow electric cars to travel for longer before having to recharge.*

◀ SOLAR POWER
*Energy from the Sun is low-cost and causes little pollution. These solar cells trap sunlight and convert it to electricity, but at the moment they are not very efficient. New materials might help us create better solar panels.*

## METAL FOAM

Foamed plastic, containing tiny bubbles of gas, has many uses, such as lightweight packaging and space-filling. Future metal foams could have similar uses, but would be stronger, fire-resistant, and more easily recycled. For instance, metal foam could be put into walls and gently heated by electricity to keep the room warm.

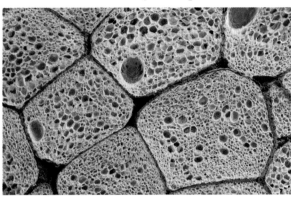

▲ ALUMINIUM FOAM *This microscope photo shows tiny bubbles in a foam of aluminium metal. This creates a very strong, but light substance that could be used in buildings.*

## STRONG BUT FLEXIBLE

Screens are everywhere – on televisions, computers, tablets, and smartphones. At the moment, screens are hard and fragile, but new materials, such as carbon-based graphene, may produce flexible touch screens that can be rolled up or even folded without breaking.

▶ BENDY DISPLAY *Flexible touch screens containing electronic circuits might mean your tablet or smartphone could be rolled up and carried in your pocket, or wrapped around your wrist.*

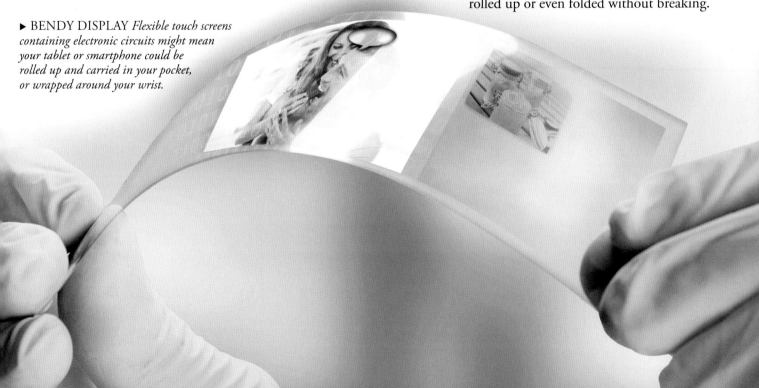

**SUPER-LIGHT SOLID**

This astonishing solid, called SEAgel, is so light that it can sit on soap bubbles without breaking them. It is made from agar, a jelly-like substance extracted from seaweed, which is dried and puffed full of gas bubbles. An apple-sized piece of SEAgel weighs the same as a single grain of rice.

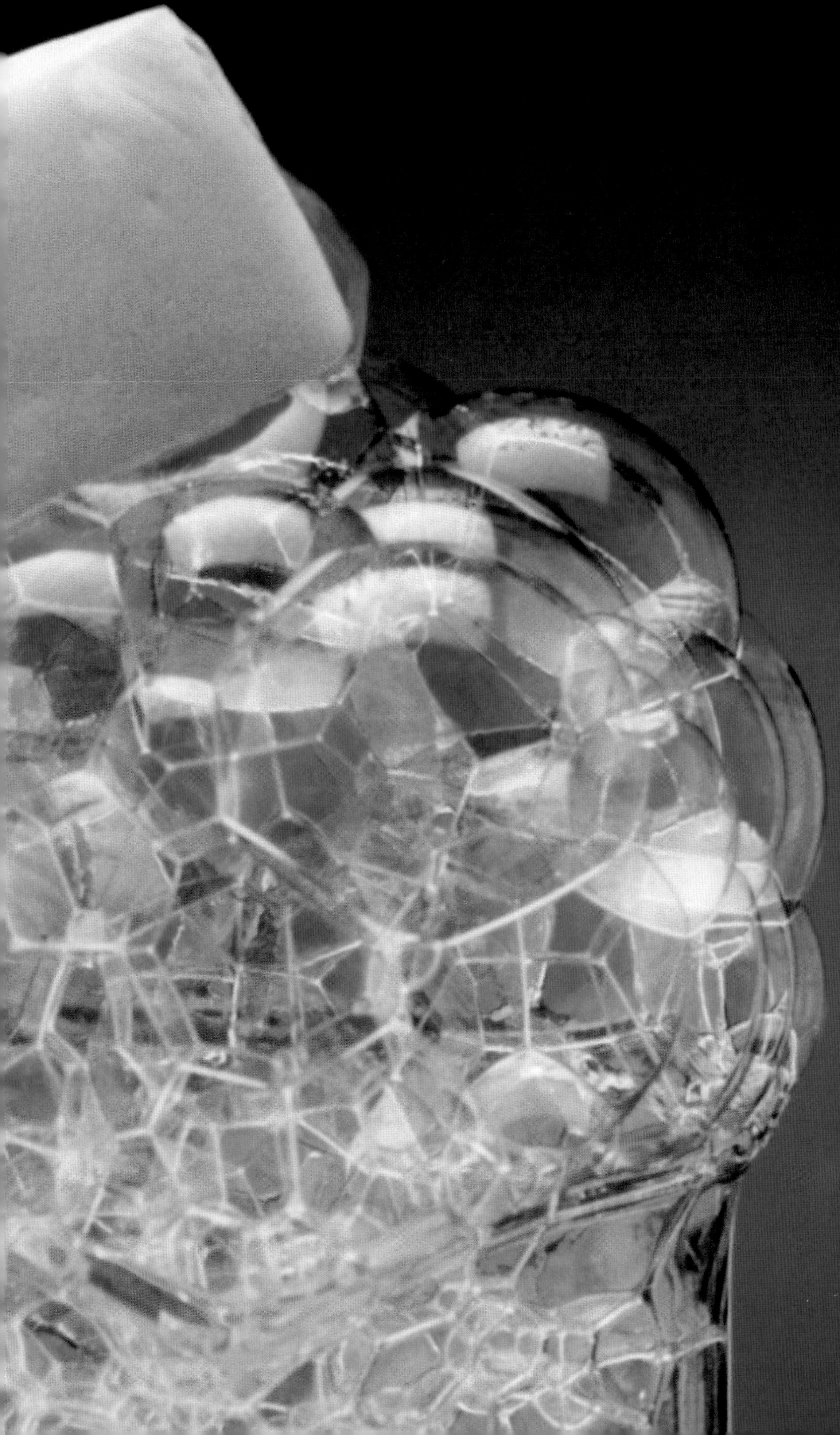

# FORCES AND MACHINES

**BALANCING ACT**
The Millau Viaduct in southern
France uses carefully balanced forces
to hold 2.5 km (1.6 miles) of road
in mid-air. Tall masts and extra-strong
cables keep the road in place so cars
can drive safely across.

A force is a push or a pull. It makes an object speed up or slow down. By building clever machines, human beings can use forces to perform tasks that our own bodies could never manage.

# What are forces?

When fireworks zoom through the sky, cars screech to a halt, or flies creep up walls, forces are at work. Forces are the hidden power behind everything on Earth – and far beyond. A force is a push or a pull, sometimes at close range and sometimes from far away. Forces can be tiny or huge. Minuscule forces glue atoms together, while the vast force of gravity hauls the planets round the Sun.

## WHAT EFFECT DOES A FORCE HAVE?

We cannot always see forces at work, but they can have spectacular effects. Often a force changes something's shape or makes it move in a new way, though forces can also hold things perfectly still. When the boxer hits this punchbag, the force makes it move, then changes its shape, and finally makes it explode. So even one force can have several different effects.

## FORCES AND YOU

There is no escaping forces. Even when you are sitting still or sleeping at night, powerful forces are acting on your body. Forces can help you walk, talk, run, and swim. Inside your body, forces help you breathe, pump blood through your veins, and even make the thoughts that ping through your head. Without forces, life would be impossible.

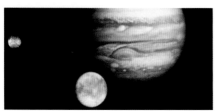

**Gravity**
*Gravity is the force that sticks you to the Earth's surface, and it also pulls the planets through space.*

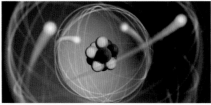

**Nuclear force**
*Every single atom inside your body is glued together by incredibly strong forces that bind its nucleus (centre) tightly together.*

**Weight**
*Weight is a measurement of the force of gravity pulling on an object. It can be measured using a spring balance.*

**Electromagnetism**
*This force, made by electrons, controls magnetism and electricity, such as the electrical signals that zap through your brain.*

## WOW!

The force of Earth's gravity reaches all the way from the centre of our planet far into space across the Solar System.

## FORCES ACTING TOGETHER

Forces do not always work alone. Instead, two or more forces can push or pull on the same thing at the same time. When this happens, the forces either add together (making a bigger force) or cancel out (making a smaller one). When forces cancel completely, it looks as though no force is acting at all.

Strong, tight steel cables pull up

Towers balanced by cables on either side

Towers reach deep underwater, anchoring bridge to seabed

Weight of deck pushes down

▲ FORCES IN BALANCE *The Golden Gate Bridge in San Francisco, USA, does not fall into the water, even though its huge deck (road) is heavy. The weight of the deck pulls down, but the steel cables pull back up. These two forces – weight and tension – cancel out so the bridge is always safe to use.*

▲ FORCES OUT OF BALANCE
*Different sized forces can act in different directions at the same time. When two wrestlers start to push against one another, they go nowhere as their pushing forces are the same. Soon one pushes a little harder, making an overall force in that direction so that the pair topple over.*

Wind blows sail, tipping boat one way

Sailor balances boat by leaning the other way

▲ FORCES COMBINED *Sometimes, many forces act at once. This boat is being blown over by the wind but the sailor is leaning out to stop it, and these forces are balanced. The wind is blowing the sail forwards, while the rough sea is pushing it backwards. These forces are out of balance, so the boat speeds onwards.*

# Forces and movement

Forces help you walk, and move your mouth when you talk. They blast rockets into space with a roar and blow ships over stormy seas. You could never run or swim without forces. With no force pumping your lungs, you could not even breathe. If there were no forces, the world would be a totally still and silent place. Nothing would move and nothing would ever happen.

## HOW FORCES MAKE MOVEMENT

Why does a rocket zoom into the sky when the force from its engine fires beneath? An English scientist, Isaac Newton, was the first person to explain how forces make things move. More than 300 years ago he worked out three simple rules, often called Newton's Laws of Motion. They are among the most important scientific concepts ever discovered.

◄ NEWTON'S FIRST LAW *Nothing moves unless a force acts on it. A rocket stays still on the launch pad until the engines fire up. If something is already moving, it keeps moving unless a force stops it.*

◄ NEWTON'S SECOND LAW *When a force pushes or pulls on something, it usually gives it extra speed. This is called accelerating. As the engines fire under a rocket, they give it speed that blasts it high into the sky.*

◄ NEWTON'S THIRD LAW *When a force pushes, there is always an opposite force of the same size pushing the other way. As the exhaust gas fires down from a rocket, the rocket fires up into the sky.*

## KEEP STILL, KEEP GOING

Heavy things are harder to move because of inertia. This means they do not like to change. A stationary truck is harder to move than a car because of its inertia. A moving object has momentum. The heavier or faster it is, the more momentum it has. A moving truck is harder to stop than a car because it has more momentum.

▼ NEWTON'S CRADLE *These balls do not move until you add force. Then they gain momentum and keep bouncing off each other until eventually they lose all their energy.*

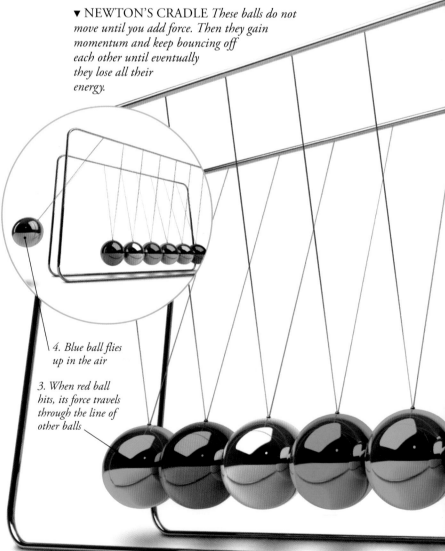

4. Blue ball flies up in the air

3. When red ball hits, its force travels through the line of other balls

## MEASURING MOTION

Speed, velocity, and acceleration are three different ways of measuring how things move. Speed is how fast something goes, which is how much distance it covers in a certain amount of time. Faster things go further each second than slower things. Velocity is the speed something has in a certain direction. If it changes direction, its velocity changes even if it its speed stays the same. Acceleration is how fast something speeds up or slows down (changes velocity).

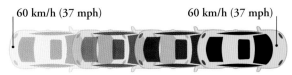

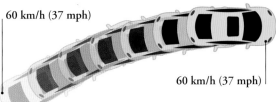

▲ SPEED *This car has a steady speed so it covers the same distance every second. If it travels for twice as long, it goes twice as far.*

▲ VELOCITY *This car goes at a steady speed but in a curve. Its direction is constantly changing, so its velocity is changing even though its speed stays the same.*

▲ ACCELERATION *This car is accelerating, which means its speed is increasing every second. The longer it travels, the faster it goes. Things accelerate when forces act on them.*

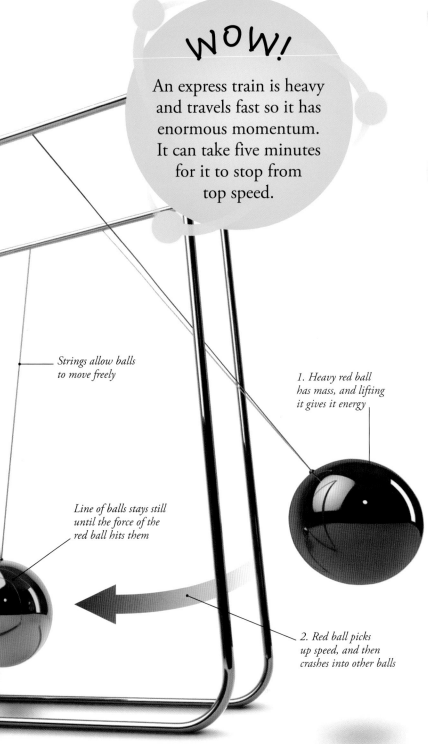

**WOW!**

An express train is heavy and travels fast so it has enormous momentum. It can take five minutes for it to stop from top speed.

*Strings allow balls to move freely*

*Line of balls stays still until the force of the red ball hits them*

*1. Heavy red ball has mass, and lifting it gives it energy*

*2. Red ball picks up speed, and then crashes into other balls*

## BRAKING FORCE

It takes force to move things – and force to stop them. The quicker you need something to stop, the bigger the force you have to use. Cars are designed to crumple when they crash so that the impact lasts longer. If a car takes twice as long to stop, the people inside feel half as much force on their bodies. That means fewer injuries and more chance of surviving.

▲ CRASH TEST DUMMIES *These dummies are used to test how the force of a crash would affect people inside the car.*

# Turning forces

Spinning, wheeling, twisting, turning – when things move in circles, hidden forces are hard at work. Left to themselves, moving things go in straight lines. It takes a force to make something turn in a circle instead. You need a bigger force to turn heavier things, spin them more quickly, or bend them in tighter circles.

## ROUND AND ROUND

The force that makes things move round in circles is called centripetal force ("centre-seeking"). On this fairground ride, the centripetal force comes from the tight sturdy ropes, which pull each person round in a different circle. If you cut the ropes, there would be no centripetal force, and the people would fly off in straight lines.

### STORING SPIN

Giant machines need to store energy so they can run smoothly. This old-fashioned steam tractor uses a huge, heavy flywheel. The engine spins the flywheel, which drives the road wheels. Even if the engine slows down, the flywheel keeps the road wheels turning at a steady speed.

▶ GOING IN CIRCLES
*The more people on the ride, and the heavier they are, the more force the wheel needs to make them spin.*

You need force to turn things in circles.

## BALANCING ACT

Your weight seems to be concentrated in your tummy, at a point called your centre of gravity. If this is not above your feet, gravity makes you topple over. When you lose balance, your body spins round your centre of gravity as you fall.

◄ PERFECT BALANCE
*This gymnast balances because her centre of gravity is above her hands. If her feet move, her body will turn and topple over.*

## STEADY SPIN

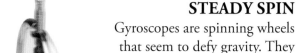

Gyroscopes are spinning wheels that seem to defy gravity. They stay upright even when you push or knock them over. A heavy wheel spinning inside the gyroscope constantly pulls it upright, so it can balance almost anywhere. Because they keep pointing the same way, gyroscopes are used like compasses to help ships and aeroplanes navigate.

*Heavy metal rim of gyroscope gives it spinning momentum*

*Gyroscope spins and balances on tiny pencil point*

**WOW!**

Standing still on the Equator, you are really moving at 1,600 km/h (1,000 mph). Earth turns that fast!

## IN A WHIRL

Hurricanes (also called cyclones or typhoons) spin around like gigantic wheels because forces act inside them. In the eye (middle), air pressure is low, creating the force that spins the hurricane. Air sucks in from the outside, making fierce winds, and the whole thing whirls round as it blows along.

► SPINNING STORM *Hurricanes look still from space, but this one is moving at 95 km/h (120 mph) – faster than the top speed of most cars.*

*The outside moves a longer distance than the inside, so it travels faster.*

*As the wheel turns, the inside moves only a short distance.*

## WHEELS AS LEVERS

Wheels work like levers and can increase force or speed. If you turn the hub (centre) of a wheel, the rim (edge) spins faster but with less force. If you turn the rim, the hub turns slower but with extra force. That is why taps sometimes have wheels fitted to help you turn them on or off.

◄ LONDON EYE, UK *This ride turns very slowly at the centre. But because the wheel is huge, the passenger cars around the edge move more quickly.*

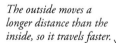

## SWIRLING STORM

Tornados form when a special type of thunderstorm creates spinning forces in the air. The biggest tornados have wind speed of more than 450 km/h (280 mph), enough to knock down a house. Less powerful is a "landspout". As seen here, this small swirling storm forms at ground level and gets pulled up into storms above.

# Friction

Every time you stride across the floor, your feet stick to it very slightly. If they did not, you would fall over. This sticking force is called friction – and it is the same force that makes your hands hot when you rub them together. Friction causes sticking and rubbing, but it also explains slipping and sliding. When there is not much friction, things slide past one another very easily. You slip on ice because there is very little frictional force.

### HANGING IN THERE
Friction can hold things still, even when other forces are pushing or pulling. Lizards can scale rocks by splaying out their toes to create friction on the rough surface. The force of friction is an upward pull, strong enough to balance the lizard's weight, pulling down.

◄ ROCK CLIMBER *Lizards can climb even vertical surfaces because their rough skin creates friction to stop them from falling.*

### TEARING AWAY
Friction holds things, but also helps them move. A car speeds along the road because its tyres grip the surface. As the wheels turn, the tyres push back on the road. The engine's power can then shoot the car forwards. Without friction, the wheels would spin, and the car would go nowhere.

*A thin layer of water underneath the surface of the skate makes it slide.*

► ICE SKATING
*The low friction between the skater's blade and the ice allows her to glide smoothly.*

## SMOOTHING THE WAY

Friction can really slow things down. It can wear out machines by slowly grinding away their moving parts. Oiling parts makes them slide smoothly past each other, so there is less rubbing. This is called lubrication, and it explains why ice is slippery. When you walk on ice, it melts a tiny bit and the water between your feet and the frozen ground makes you slide, just like oil.

## WHAT MAKES FRICTION?

Seen under a microscope, even the smoothest things are rough. If two smooth surfaces rub together, atoms on the edge of one catch and snag the atoms on the edge of the other. This snagging is one of the things that causes friction. When atoms are very close, they can attract one another electrically, and that also increases friction. The rougher the surfaces, the more friction they create. If there were no friction, things would keep on moving.

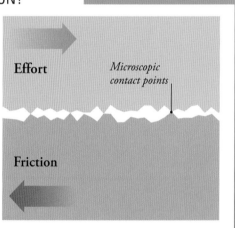

Effort

*Microscopic contact points*

Friction

◄ WHEEL SPIN
*If a car's engine makes too much force too quickly, or the tyres cannot grip, the wheels spin on the road. As the tyres rub, friction creates heat and smoke.*

## WHAT A DRAG

Air might seem empty, but if you move fast enough you can feel it pushing into you. This is called air resistance or drag, and it is a kind of friction. The faster you go, the harder you have to work against drag. Aeroplanes, cars, and even cyclists have to use lots of their energy to simply push through the air.

◄ LEAN MACHINE *Speed cyclists can go faster by wearing tight-fitting clothes and crouching down to reduce drag.*

# Gravity

Gravity holds the Universe together like a giant, invisible spider's web. Everything around you, the ground beneath your feet, and every single star and planet in space is connected to everything else by this immense tugging force. When you drop something, this same force – gravity – pulls it down towards the Earth.

FORCES AND MACHINES

## WHAT IS GRAVITY?

Gravity is the pulling force between every object and every other object in the Universe. The closer together two things are, and the heavier they are, the bigger the force of gravity between them. Earth's gravity pulls everything around it towards the centre of the Earth, keeping us on the ground.

◄ FALLING WATER
*The irresistible force of gravity pulls water downhill from mountains all the way to the sea.*

## WOW!

Tiny things have gravity just like big ones. When an apple falls to Earth, it also pulls Earth up to meet it by a tiny amount.

*Saturn has more than 60 moons around it, but when this orrery was made only five moons were known.*

## IN A SPIN

The planets zoom around the Sun in curved paths called orbits. Usually, moving things go in straight lines, but if a force pulls on them, they bend in a curve instead. Gravity from the Sun pulls the planets around it in orbits. Because each planet weighs a different amount, it moves in a different orbit. The orbits look like circles but are really oval shapes (ellipses).

► ORRERY *This antique model of the Solar System shows how the Sun's gravity pulls planets around it in orbit, and how planets pull their moons into orbit around them.*

## FREE FALL

When you jump from a plane, gravity tugs you towards the ground, but air pushes against you at the same time. The longer you fall, the faster you go. But you soon reach a speed where the force of gravity pulling down is the same as the force of air pushing up. Then you free fall – move at a steady speed as if you were weightless.

▲ EARTH *Our planet is big and heavy so the force of gravity pulling on you is strong. You cannot jump very high or far.*

▲ MOON *The Moon is smaller and lighter so the force of its gravity is weaker. You can jump much higher and further.*

*Rip-proof nylon is strong enough to survive force of air pushing up*

*Parachute ready for safe landing*

▲ SKYDIVE WING SUIT *The extra material in this suit acts like a plane's wing, pushing the skydiver up so he falls more slowly.*

*Air pressure flattens skin against face*

*Wrist computer shows flight time and direction*

## MASS AND WEIGHT

Your mass is a measurement of how much "stuff" (matter) you are made of. Your weight measures how much gravity pulls on your mass. Your mass is the same wherever you go, but your weight changes from place to place as the force of gravity changes. On the Moon, you weigh about six times less, because the Moon is smaller and less massive than Earth, so its gravity pulls less forcefully.

## MAXIMUM GRAVITY

There are spots in space where the force of gravity is so huge that even passing light gets sucked in. These odd places would look completely dark, so we call them black holes. Despite their name, they are not empty – they are packed full of mass, giving them immense gravity. If you fell towards a black hole, it would stretch your body thin, like a strand of spaghetti.

# Bending and stretching

Twisting, bending, squashing, and stretching – all these things happen because of forces. When you squeeze something soft and stretchy, such as a rubber ball, it changes shape. If you stop squeezing, you take the force away, and the ball changes back again. We say the ball is elastic, because it goes back to its original shape.

## STRETCHING SCIENCE

If you apply force to something stretchy, for example by blowing into bubble gum, it changes shape. If you blow twice as hard, it stretches twice as much. Blow too hard and it snaps. This basic rule of stretching is called Hooke's Law, after the scientist Robert Hooke.

▲ LITTLE FORCE *Chew the gum to soften it, and then blow into it. The blowing force stretches the gum to create a balloon shape.*

▲ TWICE THE FORCE *When you blow the gum with twice the original force, the balloon stretches twice as far.*

▲ TOO MUCH FORCE *If you keep blowing, the gum stretches too far and the balloon bursts.*

### BEND ME, SHAPE ME

Elastic things go back to their original shape, but not every object is like this. Many things simply bend or snap when you push them with too much force. We say they are "plastic", even if they are made of metal or another material. Plastic means things bend permanently out of shape when you push or pull them.

▶ METAL BENDING *These forks and spoons are metal, but they are also "plastic". This means their shape changes forever when you put too much force on them.*

# THE POWER OF STRETCH

Pull back on a catapult and the force you use stretches the elastic. This stores energy inside it, known as potential energy. When you let go, the elastic returns to its original shape. The stored energy must go somewhere – so it is given to the stone, flinging it through the air.

*Stone gains energy originally stored in elastic*

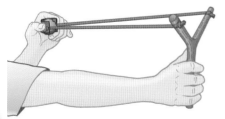

**1. Stretching elastic stores energy**

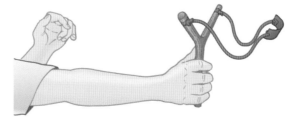

**2. Letting go releases energy**

## STRETCHY SKIN

Pinch your skin and let go and it springs straight back again. This is because young skin is very elastic. Older skin is not so elastic, which is why older people have more wrinkles. Anti-wrinkle creams work by covering skin in gluey chemicals that stretch it tight. This makes wrinkles disappear – at least for a time.

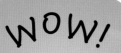

**WOW!**

Rubber can be stretched to three times its length. Hydrogels, the world's stretchiest materials, can be pulled to 20 times their original length.

## MAKING RUBBER STRONGER

Rubber comes from trees, and it is white, gooey, and smelly. This pure kind of rubber, called latex, is used to make stretchy things like balloons, but it is not very strong. We can make rubber more useful by cooking it with sulphur. This makes the rubber hard, black, and tough. It is not as stretchy, but it is stronger and lasts longer.

▲ BALLOON *The molecules inside the rubber pull far apart and spring back to shape.*

▼ TYRE *Sulphur bridges between molecules make the rubber harder to stretch, but tougher.*

*Sulphur bridge*

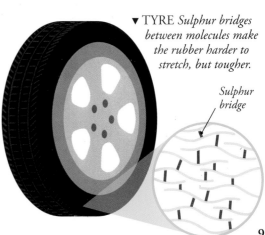

99

# Simple machines

When a giant rock blocks your path, you need the power of a digger to push it aside. Machines work by turning small forces into big ones. Even complex machines are often made by linking together smaller, simpler devices. These include levers, gears, wheels, and screws. While some machines make forces bigger, others make things go faster.

## MORE FORCE

A lever is a long rod that gives you extra pushing or pulling force. The longer the lever, the more force it gives you. A seesaw is an example of a simple lever. If you sit on the end of a seesaw, you can lift someone much heavier than you are, as long as you sit further away from the balancing point. Crowbars, hammers, and many other simple tools also work like levers.

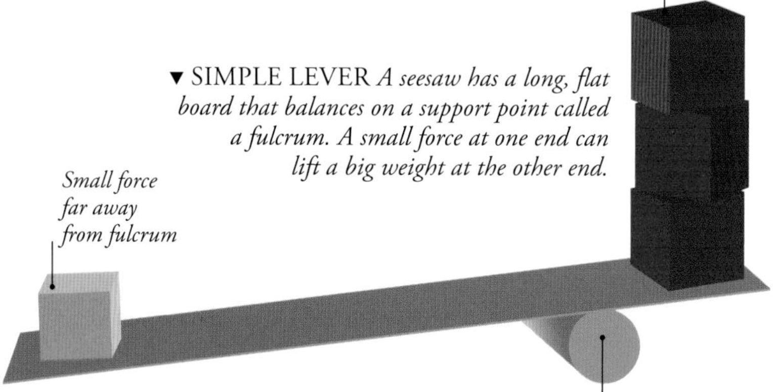

*Big force close to fulcrum*

▼ SIMPLE LEVER *A seesaw has a long, flat board that balances on a support point called a fulcrum. A small force at one end can lift a big weight at the other end.*

*Small force far away from fulcrum*

*Fulcrum (support point)*

▶ MIGHTY MACHINE
*This digger's arms use levers to lift heavy loads. Giant tyres spread the weight over a large area to stop the digger sinking in mud.*

*Levers lift arms*

## WOW!

If you had a lever big enough, you could lift the world. But the lever would need to be about 100 million trillion km long!

## MORE SPEED

Wheels are like levers that move in circles. If you turn the centre of a wheel, the outside edge has to go further to keep up, so it goes faster. If you turn the outside, the centre turns slower and with more force. Wheels can increase force or speed, but not both at the same time.

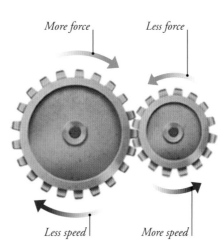

More force    Less force

Less speed    More speed

▲ GEARS *When a large gear turns a small gear connected to it, the small one spins faster but with less force.*

▲ PENNY FARTHING BICYCLE
*By pedalling the middle of the large wheel, you make the tyre turn quickly, speeding the bicycle along.*

## DIFFERENT SPEEDS

Gears are pairs of wheels that can give a machine extra force or speed. They lock and turn together using teeth around the edge to stop them slipping. If we use lots of gears, we can make different parts of a machine turn at different speeds, for example the hour, minute, and second hands on a clock.

## MORE LIFT

If you want to lift something heavy, you can use a set of wheels and ropes called a pulley. The more times the rope is wrapped around the wheels, the bigger the weight you can lift. But you have to pull the rope further to lift the weight the same distance.

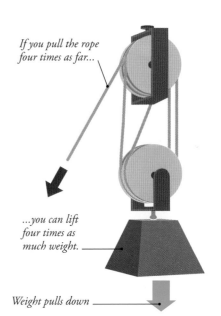

*If you pull the rope four times as far...*

*...you can lift four times as much weight.*

*Weight pulls down*

**Four-rope pulley**

## MORE PUSH

It is tough to lift boxes straight up because you have to work hard against the force of gravity. A ramp makes it easier to move heavy things uphill because you can use less force, though you have to push upwards over a longer distance.

**TUNNEL BORER**
Giant drilling machines like this one chew their way through rock. Their faces are covered in spinning wheels with teeth that grind away the stone. This borer is being used to build the Gotthard Base railway tunnel, stretching for 57 km (35 miles) under the mountains in Switzerland.

# Engines and vehicles

When you see a jet plane sweeping through the sky, or watch a car whizzing past, powerful engines are racing inside them. Most vehicles are powered by engines – machines that burn fuel to let out heat. Even tiny amounts of fuel release huge amounts of energy when they burn. An engine turns this heat energy into kinetic (movement) energy.

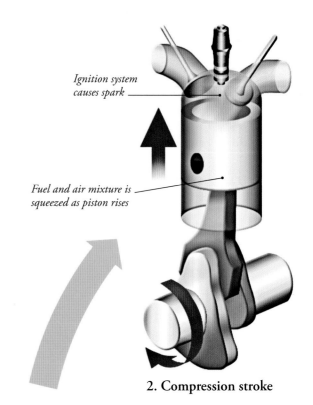

*Ignition system causes spark*

*Fuel and air mixture is squeezed as piston rises*

**2. Compression stroke**

## WHAT'S INSIDE AN ENGINE?

Engines take in fuel, burn it to release heat, and use that to make movement. In a car engine, these things happen in sturdy "cooking pots" called cylinders, with pistons at the bottom that pump up and down. As the fuel burns, each piston pumps in turn, driving a rod called the crankshaft. The spinning crankshaft carries the engine's power to the gears and wheels.

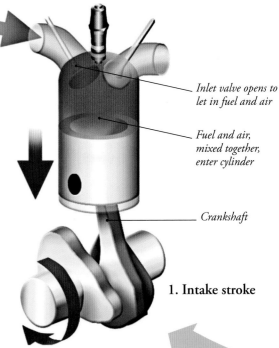

*Inlet valve opens to let in fuel and air*

*Fuel and air, mixed together, enter cylinder*

*Crankshaft*

**1. Intake stroke**

## MAKING A SPARK

In a petrol engine, a mixture of fuel and air is squeezed tightly together in the cylinder, then exploded with an electric spark from a spark plug. The explosion as the fuel burns in air powers the engine. Spark plugs work by taking power from a battery and passing it through electrodes. The sparks go off when the pistons have completely squeezed the fuel.

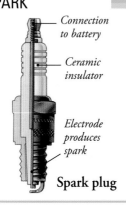

*Connection to battery*

*Ceramic insulator*

*Electrode produces spark*

**Spark plug**

▲ BURNING FUEL *Car engines repeat four steps (strokes). First, in the intake stroke, mixed air and fuel are sucked in. Second, in the compression stroke, the piston squashes the mixture. Third, in the power stroke, a spark plug makes the fuel burn, expand, and drive the piston. Finally, in the exhaust stroke, the piston pushes waste gases from the cylinder.*

## TYPES OF ENGINES

The engines on cars, planes, trains, and rockets all work in different ways. This is mainly because bigger vehicles need to make much more power than smaller ones. They have more powerful engines, so they can burn fuel more quickly, make more energy each second, and go faster.

◄ CAR ENGINE
*These run on petrol or diesel fuel. They are fairly small and light, because cars need to be economical.*

◄ JET ENGINE *These are much bigger because it takes far more energy to lift a plane and fly it at very high speed.*

◄ STEAM ENGINE
*These burn huge amounts of coal to release enough energy to pull many trucks or carriages.*

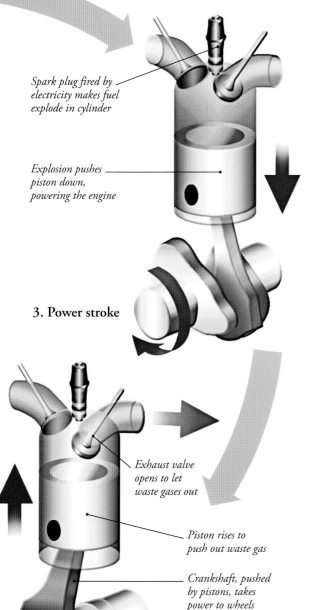

Spark plug fired by electricity makes fuel explode in cylinder

Explosion pushes piston down, powering the engine

3. Power stroke

Exhaust valve opens to let waste gases out

Piston rises to push out waste gas

Crankshaft, pushed by pistons, takes power to wheels

4. Exhaust stroke

◄ ROCKET ENGINE
*These need huge amounts of fuel to escape from Earth. They carry their own oxygen supply because there is no air in space to use for burning fuel.*

## HOW DOES AN ENGINE DRIVE A MACHINE?

Jet engines fire exhaust gases backwards, which makes a plane shoot forwards. In cars and trains, the power from the engine is used to turn the wheels. In ships (right) and small planes, the engines turn propellers, pushing air or water to power along.

# Flight

We can run and jump, but we cannot fly. No matter how hard you flap your arms, you can never make enough lifting force to overcome the pull of gravity. A plane can fly, even though it weighs 15,000 times more than you and has 500 heavy people on board. It uses powerful engines to speed forward, so its huge wings can sweep it into the sky.

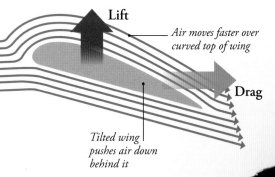

Lift

Air moves faster over curved top of wing

Drag

Tilted wing pushes air down behind it

## HOW DO WINGS WORK?

A plane's engines move it forwards, but it is the wings that make it fly. They create lift (an upward force) to overcome its weight. Wings have a curved shape that drives air down behind them, pushing the plane into the air.

## WINGS VERSUS WEIGHT

The heavier a plane is, the bigger the wings it needs. The bigger the wings, the more lifting force they can make. A typical jet plane has wings 70 m (230 ft) wide from tip to tip. That is about 60 times wider than your outstretched arms. A small plane that carries only four people has a wingspan of just 10 m (35 ft).

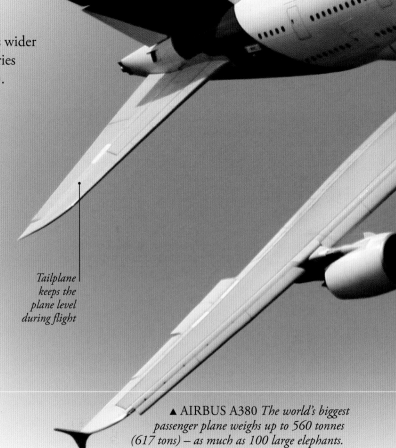

Narrow body reduces air resistance

Tailplane keeps the plane level during flight

▲ AIRBUS A380 *The world's biggest passenger plane weighs up to 560 tonnes (617 tons) – as much as 100 large elephants. That is why it needs wings 80 m (262 ft) wide.*

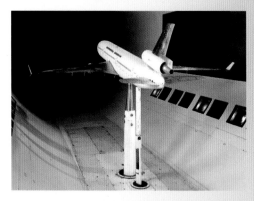

## WIND TUNNEL TESTING

It is dangerous to test a new plane. That is why exact models of planes are tested in wind tunnels. A model is fixed inside the tunnel and a fan blows air past it. This is just like flying a big plane through still air. Engineers study how air moves round the model to improve the real plane.

## DON'T TAKE OFF!

Racing cars go so fast that they could take off, which is dangerous. That is why they have low-down "wings" at the front and high-up ones at the back. They work like a plane's wings in reverse, pushing downwards to stick the car to the track.

▲ GROUND FORCE *This wind tunnel test shows how air flowing over a Formula One racing car clamps it to the ground.*

## WOW!

A Formula One racing car generates so much downward force at top speed that you could drive it upside-down on the ceiling.

## THE WRIGHT WAY

The Wright brothers changed history by making the first powered flight in 1903 – and science was the key to their success. They built their own wind tunnel from a wooden box and an old engine, and used it to test 200 different wing designs. They made around 1,000 test flights, carefully recording the results to perfect their flying machine.

**The Wright brothers' 1903 Flyer plane**

## GLIDING THROUGH AIR

Paragliders use huge parachute wings to lift themselves into the sky. The curved wing, made of very light fabric, is wide enough to generate powerful lift when the wind blows across it. By steering through upward flowing air columns, the pilot can fly for several hours before gently drifting back to Earth.

# Planes and helicopters

Gravity is the enemy of things that fly. Planes and helicopters work hard to fight gravity in two different ways. Planes use their wings to generate a lifting force, but this only works if they fly quickly through the sky. Helicopters need to hover in one place, so ordinary wings are no use to them. Instead, they generate lift using rotors (spinning wings) to pump air down beneath them.

**IGOR SIKORSKY**

When he was a boy, this ingenious Russian inventor had a dream about flying a strange machine. Forty years later, after moving to America, he built the first practical helicopter.

## POWERING PLANES

Four forces act on a plane as it flies. Thrust from the engines pushes the plane forwards, while drag (air resistance) tries to pull it back. Weight (gravity) pulls it down, while lift from the wings pushes it up. During take-off, the plane generates extra lift to climb into the sky. When it lands, it generates less lift so gravity brings it back down again.

Lift

Thrust

Drag

Weight

◀ FORCES ON A PLANE
*A plane flies forwards because the thrust from its engines is always bigger than the drag.*

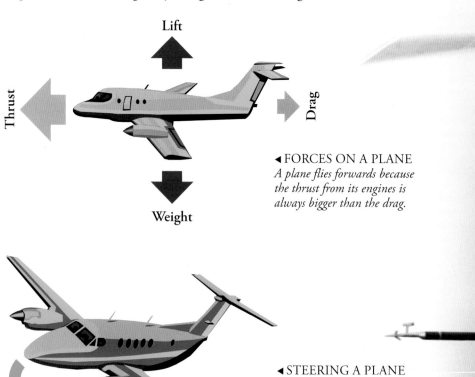

◀ STEERING A PLANE
*Planes roll left or right by using small moving surfaces on the wings called ailerons. Surfaces on the tail are used to point the aircraft up, down, left, or right.*

More lift on one side steers the plane to the other side.

*Probe for refuelling in mid-air*

## DI 147a    KUM☺N

Statements from
Paragraphs 5

Name: _____

Date:        /        /

Time:        :    —    :

**I** Read the sentences. Then choose and write the word
from each sentence which matches the definition. (You
may answer with the same form of the word as appears in
the sentence.)                                                    [-5 each]

▶ Our team defeated all the others to win the cup.

▶ "I'm just not very good," James said defeatedly.

| definition | word |
|---|---|
| won a victory over (verb) | 1) |
| in the manner of someone who has lost (adverb) | 2) |

▶ Dorothy was always encouraged by her friends' cheers
when she played in hockey matches.

▶ Dorothy's parents discouraged her from watching television.

| definition | word |
|---|---|
| tried to make someone lose confidence in or enthusiasm for something (verb) | 3) |
| gave support to (verb) | 4) |

▶ Our opponents were the team from Ashcroft School.

▶ Jane's parents opposed the marriage.

| definition | word |
|---|---|
| people playing against you in a contest (noun) | 5) |
| disagreed with and tried to stop (verb) | 6) |

# DI 147b

**II** Complete each sentence using the word in the brackets together with **your own words**. (You may change the form of the word in the brackets.)                    [-5 each]

1) [ opponent ] + your own words

The Ravens School team, who _____

_____ that day, were the best team in the area.

2) [ defeat ] + your own words

The Ravens School team, which _____

_____, were expected to win the

cup.

3) [ encourage ] + your own words

John, who was _____

_____, scored the winning goal that day.

## HOVERING HELICOPTERS

A helicopter's giant spinning rotor generates enough lift to balance its weight. The pilot can create more or less lift by tilting the blades at different angles. If the lift is more than the weight, the helicopter climbs. If the lift is less than the weight, the helicopter moves slowly down.

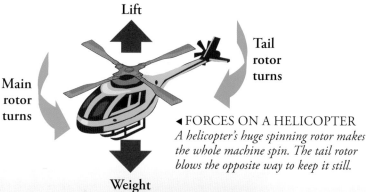

**Lift**

**Tail rotor turns**

**Main rotor turns**

**Weight**

◀ FORCES ON A HELICOPTER
*A helicopter's huge spinning rotor makes the whole machine spin. The tail rotor blows the opposite way to keep it still.*

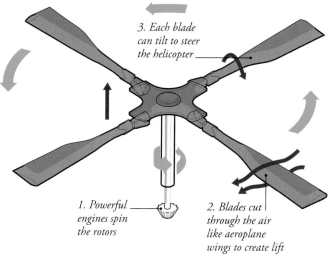

*3. Each blade can tilt to steer the helicopter*

*1. Powerful engines spin the rotors*

*2. Blades cut through the air like aeroplane wings to create lift*

▲ STEERING A HELICOPTER
*The rotor blades can tilt as they turn, creating more lift on one side. This steers the helicopter through the air.*

## CONVERTIBLE PLANE

Planes can fly fast, but need runways and cannot hover on the spot. Helicopters can hover and take off vertically, but cannot fly fast. The Osprey can do both, with giant rotors that swivel between upright and front-facing positions.

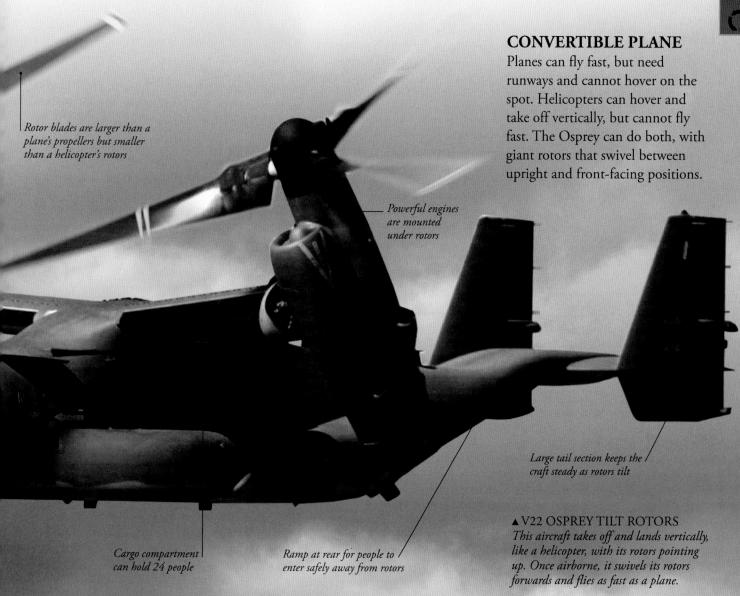

*Rotor blades are larger than a plane's propellers but smaller than a helicopter's rotors*

*Powerful engines are mounted under rotors*

*Large tail section keeps the craft steady as rotors tilt*

*Cargo compartment can hold 24 people*

*Ramp at rear for people to enter safely away from rotors*

▲ V22 OSPREY TILT ROTORS
*This aircraft takes off and lands vertically, like a helicopter, with its rotors pointing up. Once airborne, it swivels its rotors forwards and flies as fast as a plane.*

# Rockets and space flight

The quickest way to escape the pull of Earth's gravity is to climb into a giant firework – a space rocket – and aim for space. Scientists first dreamed of racing into space in the early 20th century, but it was only in 1961 that rockets became safe enough to carry people. In future, rockets will work like planes, so more of us will feel the thrill of blasting into space.

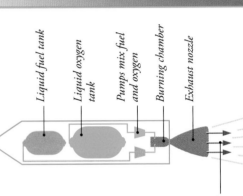

## SPEEDING INTO SPACE

Rockets have to go fast to climb out of Earth's gravity. The speed they need is called escape velocity, and it is about 40,000 km/h (25,000 mph) or 125 times faster than a racing car. A rocket needs speed because it takes energy to push against the force of gravity. A fast rocket has energy to escape, a slow one does not.

## HOW DOES A SPACE ROCKET WORK?

A rocket works like a jet engine on an aeroplane. It blasts fiery gas backwards through a nozzle to make the whole rocket shoot forwards. A jet engine makes the gas by burning fuel in oxygen (from the air). There is no air in space, so a rocket has to carry its own oxygen inside a huge tank.

Liquid fuel tank

Liquid oxygen tank

Pumps mix fuel and oxygen

Burning chamber

Exhaust nozzle

Hot exhaust gas blasts rocket upwards

## EYES IN SPACE

One of the most common uses for rockets is launching satellites into orbit. These space machines can act as mirrors to bounce telephone and TV signals across the world. Or we can use them as "eyes" in space for making maps of Earth. If we point them out into space, they can take astonishing photos of distant stars we could never capture from Earth.

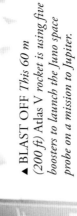

▲ BLAST OFF *This 60 m (200 ft) Atlas V rocket is using five boosters to launch the Juno space probe on a mission to Jupiter.*

WOW!

You can cry in space, but there is no gravity to make tears run. They just fill up your eyes instead.

FORCES AND MACHINES

## LIFE IN SPACE

It takes only eight minutes for a rocket to reach space, but it is usually three days before astronauts get to their final destination. Space missions can last anything from a few weeks to a year or more. While in space, astronauts mostly carry out scientific experiments. But they also have to do everyday things we do here on Earth. They need to sleep, eat, shower, and exercise – all inside a space rocket, in zero gravity (weightlessness). Rockets are carefully designed so these things are safe and easy to do in a very small area.

▲ SPACE SNOOZE *Astronauts have sleeping bags tied to the wall to stop them floating away.*

▲ GOING NOWHERE *With no room to go jogging, astronauts have to exercise on a treadmill.*

## SPACE PLANES

Rockets are great for shooting things into space, but not so good at bringing them back. When spacecraft return, Earth's gravity makes them go very fast, and friction can burn them up. In the future, rocket planes will zoom into space and come back safely. One day, they might take you on a space holiday.

▲ SPACE VISITORS *This aircraft, SpaceShipTwo, is designed to fly up to the very edge of our atmosphere, to give space tourists a taste of space flight.*

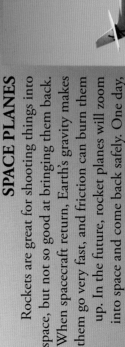

# Under pressure

When things push against each other, they create pressure. Pressure is a measure of how much force pushes against each point on an object's surface. Here on Earth, we are under constant pressure. Although you never notice, there is a thick blanket of about 100 km (60 miles) of air constantly squashing down on you. Air might seem like emptiness, but it still has mass, so gravity pulls it down towards the ground. This makes air pressure, a force spread right across your body.

## WHAT IS PRESSURE?

Pressure is what happens when a force pushes on a surface. More force makes more pressure. When the same force presses over a smaller area, the pressure becomes greater. But if the force is spread over a larger area, the pressure reduces. Sometimes, we do not notice pressure until the force is released.

◄ JET PRESSURE
*When fire fighters take the cap off a hydrant, the water inside blasts out because it is under high pressure.*

## SPREADING THE LOAD

A person can easily stand on a bed of nails. Their body has weight, which means gravity pulls it down with a lot of force. But when they stand on hundreds of nails, their weight is shared across them all. The pressure on each nail is very small so they do not get injured.

## PRESS FOR SUCCESS

Life would be impossible without pressure. Blood flows around your body because your heart pumps it with enough pressure to reach your fingers and toes. Water can flow to your home because it is stored high up in reservoirs and tanks. Gravity pulls the water down, giving it pressure that makes it spurt from the tap. Pressure is also used to make many tools work, from vacuum cleaners to drawing pins, and car engines to aeroplanes.

▲ DRAWING PINS
*A gentle pressure on the wide head creates a lot of pressure on the narrow tip of the pin.*

▲ PNEUMATIC DRILLS
*High-pressure air flowing through the hose bangs the drill into the road.*

▲ WALKING ON WATER
*Pond-skating insects spread their weight, creating little pressure so they do not sink.*

Air pressure is created by the weight of air

*Hose supplies
oxygen for
pilot to
breathe*

## WHAT IS AIR PRESSURE?

Air pressure is created by the weight of air
above you. If you climb a mountain, there
is less air above you, and therefore less air
pressure. It is harder for air to get into your body
and harder for you to breathe. High in the sky,
there is hardly any air pressure and breathing is almost
impossible. Planes have their compartments pressurized
by pumps so people inside can breathe normally.

◄ PRESSURE SUIT *Fighter pilots feel huge forces.
They wear special suits with air pumped into them.
This added pressure helps their blood flow normally.*

# WOW!

Marianas Trench is the
deepest part of the world's
oceans. The water pressure
there is 1,000 times higher
than the air pressure
on land.

## HOW TO MEASURE AIR PRESSURE?

Air pressure changes our weather.
Low pressure brings storms and
rain. High pressure means sunshine.
We can forecast the weather using a
barometer to measure air pressure.
Inside this one is a box filled with
air. As the air pressure changes, the
box squeezes in and out. This turns
the pointer around the scale.

*Pressure increases
down the carton*

*Pressure is
low, so liquid
trickles out*

*Pressure is
greater, so liquid
goes further*

*Pressure is
high, so liquid
spurts out*

**Liquid pressure in a carton**

## HEAVY WATER

The deeper you dive beneath the sea, the
more pressure there is. That is because there
is more water over your head pushing
down. Water is denser than air – the same
amount of it weighs more. So water pressure
affects things more than air pressure. That
is why scuba divers can go down only a
short distance, and why submarines need
hulls made of strong metal to withstand
the pressure. In a drinks carton, the
liquid near the bottom is squeezed by
the weight of the liquid on top.

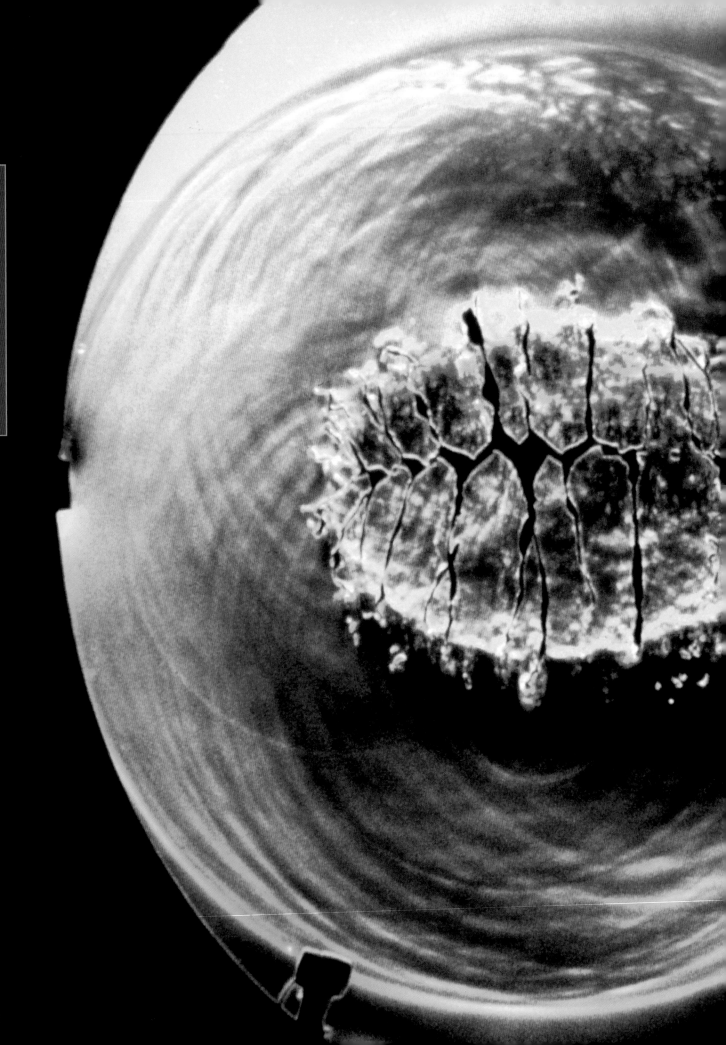

**PRESSURE WAVES**

Using a special technique, this photograph captures pressure changes in the air caused by a balloon bursting. The balloon is filled with gas from a canister. The pressure builds up inside until the rubber weakens and the balloon pops. A shockwave of pressure spreads out, which we hear as a bang.

# Floating and sinking

A blue whale can weigh as much as 20 elephants, yet it has no problem floating in the sea. The giant ships that ferry goods around the world can hold thousands of huge containers without sinking. Things float if they are less dense than the material around them – that is, if a certain volume of the object weighs less than the same volume of its surroundings.

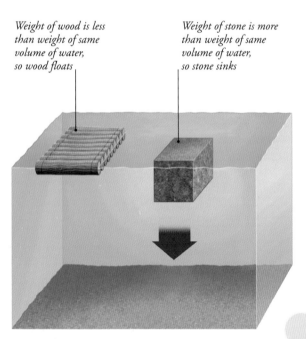

*Weight of wood is less than weight of same volume of water, so wood floats*

*Weight of stone is more than weight of same volume of water, so stone sinks*

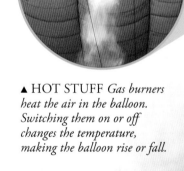

▲ HOT STUFF *Gas burners heat the air in the balloon. Switching them on or off changes the temperature, making the balloon rise or fall.*

## WHY DO THINGS SINK?

Gravity pulls things down, even in water. But water pushes upwards against things floating in it. If there is enough water pressure beneath something to balance its weight, it floats. Wood is less dense than water. The weight of a block of wood is less than the pressure of the water underneath, so it floats. Stone is more dense than water, so its weight pulls it down in spite of the pressure of water underneath.

## WOW!

In 2005, Indian Vijaypat Singhania reached a record height in a hot-air balloon of 21,000 m (69,000 ft) – twice as high as jets fly!

## GOING UP

Heavy things can fly if they can create enough upward force to lift their weight. Before planes were invented, people took to the skies in balloons. These soar into the sky when hot air is pumped inside the huge fabric dome. A balloon full of hot air weighs less than the same balloon full of cold air. It is less dense than the air around it, so it soars into the sky. Big balloons can create enough force to carry people with them.

*Envelope (dome) of balloon is made from long strips of rip-proof nylon*

## THAT SINKING FEELING

A floating ship sits partly in and partly out of the water. The heavier it is, the deeper it sits – squeezing the water underneath. It stops sinking when the water pressure underneath equals its weight. The squeezed water has to go somewhere so, when a ship is loaded up, it pushes the water aside (displaces it). The amount of water displaced always weighs the same as the ship and its cargo.

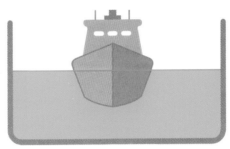

A floating ship pushes aside (displaces) its own weight of water.

The volume of water displaced weighs the same as the ship.

## CRAFTY RAFTS

The world's biggest cargo ships carry up to 18,000 vast containers, each as big as a truck. Ships work by spreading their weight over a wide area. Most of a ship is just empty space, so it weighs less than the same volume of water. Although it is incredibly heavy, it still floats.

▶ PLIMSOLL LINE *A cargo ship has an upright ruler drawn on the side. As the cargo is loaded, the ship sinks slightly and the water line on the ruler shows if it is safe to travel.*

# Boats and submarines

More than two-thirds of Earth's surface is covered by water, so getting around our planet often means travelling by boat. Boats float on the surface of the water using forces such as wind to move across it. Submarines can change their density to float or sink at will.

## HOW DOES A SHIP WORK?

A sailing ship is a mighty machine built for crossing the oceans. It has to survive four huge forces pounding from different directions. The ship's weight, and that of the cargo and crew, push down from above. What stops the ship sinking is buoyancy – water pressing up from below. Wind buffets the sails, driving the boat from behind, but the rough waves drag against the boat, pushing it back again.

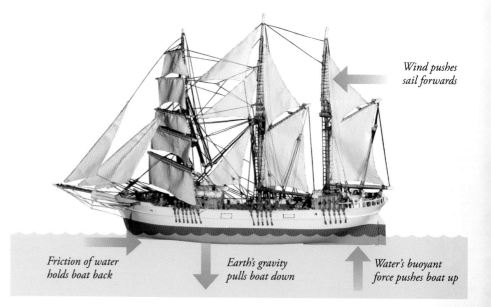

*Wind pushes sail forwards*

*Friction of water holds boat back*

*Earth's gravity pulls boat down*

*Water's buoyant force pushes boat up*

## WATER POWER

Boats do not simply glide across the waves. Their weight makes them sink under the surface a little bit, so they drag in the water as they move. That is why it can take a lot of force to get boats going. The more cargo they carry, the deeper they sink, and the harder it is for them to move. Different types of boat use different sources of power to travel across the waves.

**Sailing boats**
*Wind pushes the sails to move the boat. By angling the sails, you can travel parallel to or even against the wind.*

**Oar power**
*If you pull the oars backwards, the boat moves forwards. This is Newton's third law of motion (p.88) in action.*

**Jet skis**
*These water motorbikes squirt a powerful jet of water behind them. This pushes them forwards at very high speed.*

**Propellers**
*Most boats use propellers driven by outboard motors. The propellers pull against the water to drive the boat forwards.*

**Hovercraft**
*Hovercraft use fans to lift above the waves on a cushion of air. Another fan blows them forwards at speed.*

## RISING ABOVE

A boat can go faster if it lifts above the waves. Most boats have curved fronts so they rise up as they zip along. Hydrofoils use underwater wings. As they pick up speed, the wings lift the whole boat clear of the waves beneath. There is less drag in the water so the boat moves faster.

► DEEP FLIGHT AVIATOR
*This high-performance submarine resembles an aeroplane. It can reach a depth of 458 m (1,500 ft) under water.*

## FLYING UNDERWATER

Planes can fly because fast-moving air glides over their wings and sweeps them into the sky. This submarine, called Deep Flight, does the same thing underwater. It has side wings that push it down as water flows over them. The submarine's propellers are battery-powered, so it is very quiet. It has been used for science exploration and making films about the oceans.

*Vertical fins steer craft to left or right*

*Glass domes give all-round view*

*Propellers on each side power craft forwards*

*Large side wings make craft dive or climb*

*Narrow nose reduces water resistance*

## SNEAKING BELOW

Submarines rise and fall by changing their weight. They have giant tanks inside and pump sea water into them when they want to dive. When they want to rise again, they pump air into the tanks instead.

**On surface**
*The submarine floats by completely filling its tanks with air.*

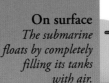

**Diving down**
*With a little water in the tanks, the submarine starts to sink.*

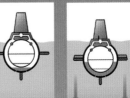

**Underwater**
*When the tanks are filled with water, the submarine sinks to its lowest depth.*

**Rising up**
*Pumping air into the tanks empties the water, making the submarine rise.*

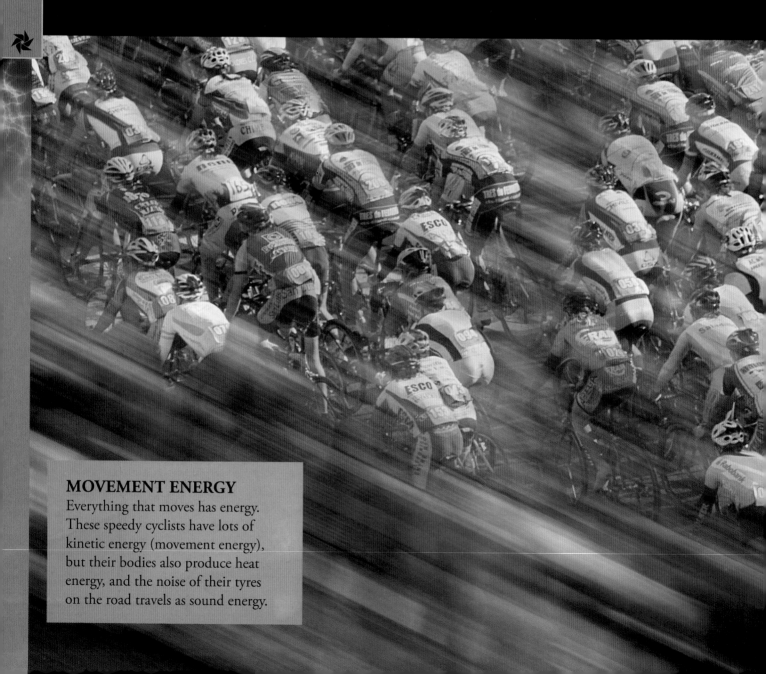

# ENERGY

## MOVEMENT ENERGY

Everything that moves has energy. These speedy cyclists have lots of kinetic energy (movement energy), but their bodies also produce heat energy, and the noise of their tyres on the road travels as sound energy.

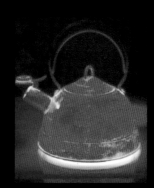

Energy is what makes things happen. Whenever something moves, gives off heat or light, makes a noise, or creates an electric current, energy is being released.

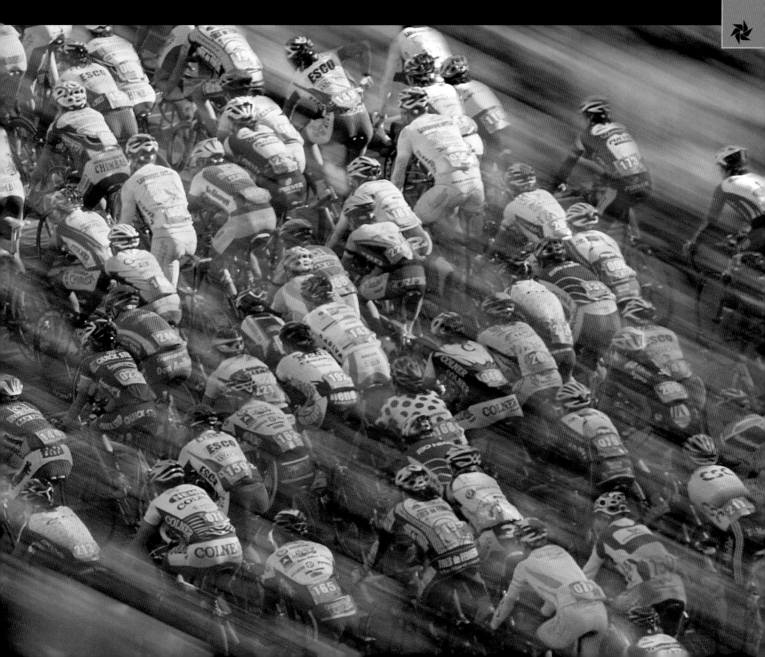

# What is energy?

Energy is what makes things happen. It is the bang behind a firework and the roar of a racing car. It pushes planets around the Sun and helps trees reach for the sky. It makes earthquakes that can tear the ground beneath our feet and music that makes us get up and dance. Energy is the hidden power behind everything – including life on Earth and all that happens here.

*Big Bang*   *Energy starts turning into matter*

## HOW ENERGY WAS BORN

The Universe began about 14 billion years ago with a gigantic explosion called the Big Bang. Explosions normally destroy things, but this one created everything we know. It was the beginning of space and the start of time. The Universe was made entirely of energy to begin with. But within a fraction of a second after the Big Bang, this energy started turning into particles of matter, then into atoms from which stars and planets are formed.

## WOW!

After the Big Bang, the Universe was incredibly hot. It took 400,000 years for it to cool enough to form the very first atoms.

## WHAT DOES ENERGY DO FOR US?

Everything you do needs energy. All the things you use are made using energy. And everywhere you go, you need energy to get there. It is no wonder that the world uses so much energy. Around 80–90 per cent of it comes from fossil fuels such as oil, coal, and gas. We burn them to release the energy locked inside, which we can then use for heating, cooking, or generating electricity. Most of the world's energy is used in three ways – in our homes, for transport, and in business and industry.

▲ ENERGY AT HOME
*Surprisingly, not much energy is used in people's homes – only about 15 per cent. Most of that goes into heating, lighting, cooking, and air-conditioning.*

▲ ENERGY FOR TRANSPORT
*About a quarter of all energy is used to power vehicles, such as cars, trucks, and trains. Most vehicles run on fuels made from oil.*

▲ ENERGY FOR INDUSTRY
*Most of the world's energy is used by industry and business. A lot goes to factories, though a growing amount is used in offices.*

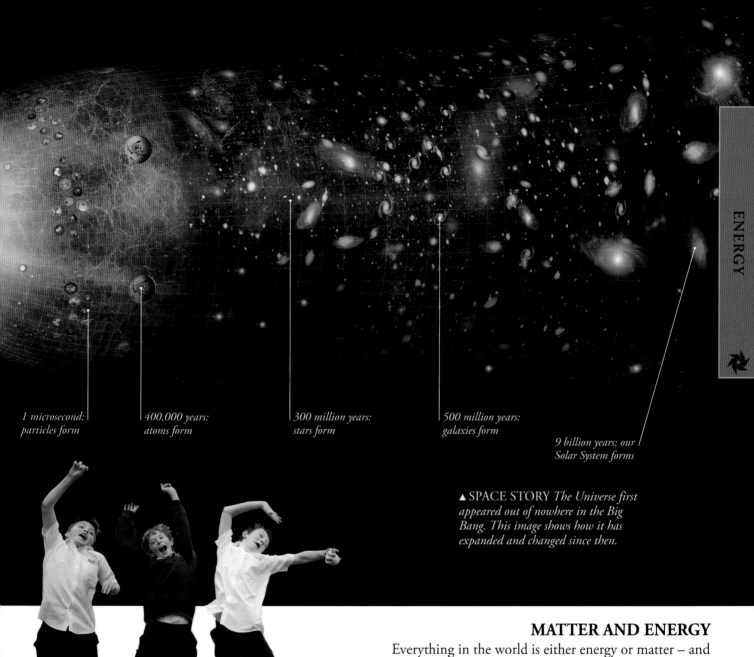

1 microsecond:
particles form

400,000 years:
atoms form

300 million years:
stars form

500 million years:
galaxies form

9 billion years: our
Solar System forms

▲ SPACE STORY *The Universe first appeared out of nowhere in the Big Bang. This image shows how it has expanded and changed since then.*

## MATTER AND ENERGY

Everything in the world is either energy or matter – and even matter is a kind of energy. Trees capture the Sun's energy in matter by using it to power the growth of their branches and leaves. If you burn wood, you release the trapped energy as heat and light.

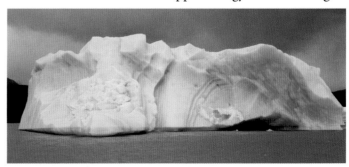

▲ ICEBERG *Even something still and frozen contains energy. There is about 200 million times more heat energy in a typical iceberg than in a cup of boiling hot coffee, but spread over a much larger volume.*

## ENERGY IN YOUR BODY

You are a living machine powered by energy. You load energy into your body by eating food, and use that energy to move, think, sleep, and breathe. Although you might think your muscles do all the hard work, a lot is done by your brain. This small bag of thoughts and feelings uses up to a fifth of all your energy, even while you are asleep and dreaming.

# Types of energy

The world is bursting with energy. Your TV flickers with light, your voice makes sound ripple through the air, and your muscles have energy packed inside them from the food you eat. A spider climbing your wall stores energy by working against the force of gravity. Even this book has energy hiding in its atoms. Everything you can see is either storing energy or using it.

## STORED ENERGY

Stored energy is called potential energy. It is like money in the bank. It is energy we have "saved up" that we can use later. We can store potential energy in different ways by working against forces. If you push a car uphill, your body works against the force of gravity, which is trying to pull the car back down. You lose energy from your muscles but the car gains potential energy. If you let go, it has the potential (ability) to roll down again.

▲ CHEMICAL ENERGY *When chemicals react together, they can give off energy. Emergency flares use stored chemical energy to make coloured smoke, signalling that a boat is in distress.*

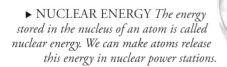

▶ NUCLEAR ENERGY *The energy stored in the nucleus of an atom is called nuclear energy. We can make atoms release this energy in nuclear power stations.*

◀ ELASTIC ENERGY *You need force to change something's shape. Springs, rubber bands, and bows store elastic energy when you stretch them and release that energy when you let go.*

## USING ENERGY

We can use stored (potential) energy to do all kinds of useful things. If you let go of a ball at the top of a hill, it rolls down. The potential energy it had at the top is quickly converted into movement (kinetic energy) as it speeds along. In just the same way, the chemical, nuclear, or elastic energy stored inside things can be changed into other kinds of useful energy.

► LIGHT ENERGY
*This luminous watch glows in the dark because its painted hands give off light. They are turning stored chemical energy into another kind of energy – light.*

◄ SOUND ENERGY *If you bang a drum, your moving hands give it elastic energy. As the stretched drumskin wobbles about, it shakes the air and the elastic energy turns to sound energy.*

## WOW!

There is as much energy in the Universe today as there was 14 billion years ago when it first came into being.

◄ KINETIC ENERGY
*This galloping cheetah is using its muscles to change chemical energy into movement (kinetic energy). The bigger things are and the faster they move, the more kinetic energy they have.*

◄ HEAT ENERGY
*Hot things have energy because the atoms inside are crashing into one another. The hotter they are, the faster the atoms move, and the more they collide.*

*Bubbling hot water has kinetic energy and heat energy.*

125

# Potential and kinetic energy

Our hectic world is powered by energy – nothing can happen without it. But where do the things that rush around us get their energy from? Things that are moving have kinetic energy, which they have to get from somewhere. Often they make it from their own stored-up energy, which is called potential energy.

## RIDING THE ROLLER COASTER

A roller coaster starts high in the air and swoops up and down as it twists and turns. This thrilling, racy ride is also a very clever energy-converting machine. When the cars are high in the air, they have a lot of potential energy. That is stored energy, ready and waiting to do things. As the cars clatter down the track, they convert some of this stored energy into kinetic energy. That is the energy things have when they move.

▶ ENERGY RIDE
*The more people there are on the ride, the more potential energy it takes to lift them into the air.*

## HOW A ROLLER COASTER CONVERTS ENERGY

High above the ground, the roller coaster has potential energy as Earth's gravity pulls down on it. When the car rolls downhill, it converts some of this potential energy to movement (kinetic energy). But the car slowly loses its energy due to friction (rubbing against the track) and air resistance. It stops when its energy is gone.

*At the start of the ride, the car has maximum potential energy.*

*The car goes fastest in dips when the potential energy turns to kinetic energy.*

*The car is always losing energy to friction and drag so it cannot climb as high.*

*The car comes to a stop when it has no more energy left to push it along.*

ENERGY

## WHAT HAPPENS TO THE ENERGY

We cannot make energy appear or disappear like magic, but we can change its form. When this athlete lifts her weight, her muscles use up stored potential energy. Her body loses energy as she lifts and the weight gains potential energy instead. When she drops the weight to the ground, its potential energy turns to kinetic energy. As the weight hits the floor, its kinetic energy turns into other types of energy – noise and elastic energy as it bounces up and down.

*The weight gains potential energy as it rises into the air.*

*Chemical energy is lost from the muscles as they push against the force of gravity.*

*The weight gains kinetic energy as it falls.*

*Kinetic energy is turned to sound as the weight hits the ground.*

*Kinetic energy is lost to the ground as the weight bounces off.*

POTENTIAL ENERGY *The athlete uses stored energy from food to give the weight potential energy.*

KINETIC ENERGY *The falling weight speeds up as potential energy is turned into kinetic energy.*

Roller coasters use science to make you scream.

# Energy spectrum

Light is energy we can see. But it is only one small part of a whole spectrum of energy rays that stream across the Universe. There are many types of energy similar to light that are invisible to us. Like light, these are made from waves of electricity and magnetism, racing around at high speed. We call these different energy waves the electromagnetic spectrum.

**WOW!**

Bees and butterflies can see colours we cannot. Their eyes can detect ultraviolet radiation that is invisible to us.

## ON THE WAVELENGTH

If you could take electromagnetic waves and stretch or squeeze them to make them longer or shorter, you could make any type of electromagnetism. The waves that carry radio and TV broadcasts are longer than a football field, while those that make gamma rays are much smaller than atoms.

◀ MICROWAVES *Microwave radiation is absorbed by water, where it turns into heat energy. Microwave ovens use this energy to heat up food.*

| Radio waves | | Microwaves |
|---|---|---|

1 km    100 m    10 m    1 m    50 cm    1 cm    1 mm    0.1 mm

◀ RADIO WAVES *TV and radio programmes race through the sky, riding on very long electromagnetic waves called radio waves. We can see distant light rays using a telescope, but to detect radio waves we need a metal aerial or a dish-shaped radio telescope like this.*

▲ MOBILE PHONE MASTS *Mobile phones use bursts of electromagnetic radiation to send messages. They use similar wavelengths to microwave ovens, but use much less energy. The signals are picked up by towers like this one.*

# ENERGY RAYS

The night sky gleams with stars – explosions of energy blasting through space. Much of the electromagnetic energy that surrounds us started off in space. Some of it, called cosmic microwave background, was originally produced during the Big Bang almost 14 billion years ago. It is the oldest light in the Universe.

▶ LIGHT YEARS *Stars are vast distances away, measured in light years. A light year is how far light travels in one year: 9.5 trillion km (6 trillion miles).*

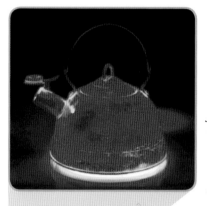

◀ INFRARED RAYS *The heat you can feel beaming from things is a kind of invisible "hot light" called infrared radiation. Like light, it travels instantly and in straight lines.*

◀ GAMMA RAYS *Some atoms give off powerful and dangerous radiation called gamma rays. They are the most energetic form of electromagnetic radiation and they can travel right across the Universe.*

| Infrared rays | Visible | Ultraviolet | X-rays | Gamma rays |
|---|---|---|---|---|

| 0.01 mm | 780 nm | 380 nm | 10 nm | 1 nm | 0.1 nm | 0.01 nm | 0.000001 nm |
|---|---|---|---|---|---|---|---|

nm = nanometre
(1 billionth of a metre)

◀ ULTRAVIOLET *The Sun pumps out light we cannot see, including harmful ultraviolet that can cause sunburn and skin cancer. Sunscreens stop this harmful radiation from reaching your skin.*

◀ X-RAYS *Light cannot get through skin or bone. More energetic X-rays get through skin but not bone. They can make "shadow" pictures of our skeletons that show up broken bones or some illnesses.*

## THE SPEED OF LIGHT

Electromagnetic waves (including light) are the fastest things in the Universe. They travel 300,000 km (186,000 miles) per second – quick enough to whip around the world 450 times in a minute.

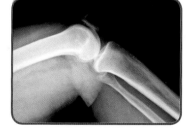

| JOURNEY | DISTANCE | TIME FOR LIGHT TO TRAVEL |
|---|---|---|
| From this book to you | 30 cm | 1 nanosecond (billionth of a second) |
| Around the Earth | 40,000 km | 0.134 seconds |
| The Moon to Earth | 384,000 km | 1.3 seconds |
| The Sun to Earth | 150 million km | 8 minutes 20 seconds |
| Nearest star to Earth | 40 trillion km | 4.2 years |
| Distance across our galaxy | 950,000 trillion km | 100,000 years |

## STAR DUST

Electromagnetic radiation from distant stars speeds through empty space. Some of it reaches Earth as visible light, while some travels in wavelengths we cannot see. This image shows infrared radiation from the North America Nebula, a huge cloud of star dust about 1,500 light years away.

# Heat

Atoms are constantly jiggling about with energy. We call this heat energy. The more heat something has, the faster its atoms move. Even the coldest objects have some heat energy as their atoms are always moving around. If we could stop the atoms moving, we could make an object completely cold. We call this temperature "absolute zero", though no one has ever managed to cool anything down that much. The Sun is probably the hottest thing you can imagine. If you could peek inside, you would see it exploding with heat.

WOW!

The centre of the Sun is about 10,000 times hotter than the temperature of melting steel.

## MEASURING HEAT

The temperature of an object tells us how hot or cold it is. We can use this to tell whether food is properly cooked, if a person is suffering from a fever, or if a car engine is overheating. These days we have electronic thermometers that can measure temperature for us quickly and accurately, but in the past humans relied on changes in nature to measure temperature.

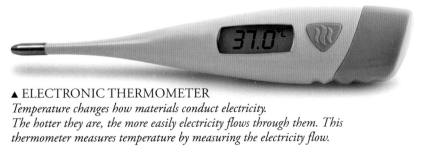

▲ ELECTRONIC THERMOMETER
*Temperature changes how materials conduct electricity. The hotter they are, the more easily electricity flows through them. This thermometer measures temperature by measuring the electricity flow.*

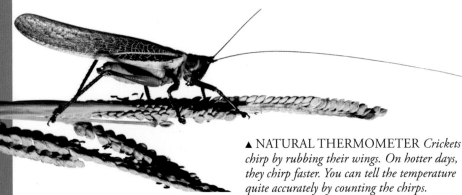

▲ NATURAL THERMOMETER *Crickets chirp by rubbing their wings. On hotter days, they chirp faster. You can tell the temperature quite accurately by counting the chirps.*

## TEMPERATURE SCALES

We can use different scales to measure temperature. In 1724, German scientist Daniel Fahrenheit invented the Fahrenheit scale (°F). He set 0°F as the lowest temperature he could create in his laboratory, and 100°F as human body heat. Today scientists use the Celsius scale (°C), with 0°C the freezing point of water and 100°C its boiling point.

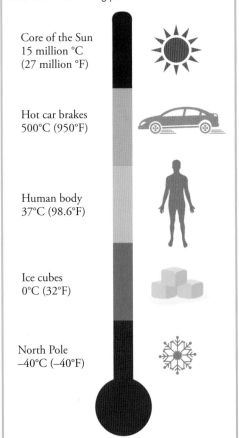

Core of the Sun
15 million °C
(27 million °F)

Hot car brakes
500°C (950°F)

Human body
37°C (98.6°F)

Ice cubes
0°C (32°F)

North Pole
−40°C (−40°F)

# HEAT CHANGES THINGS

Heat is a type of potential energy – it is stored energy that has the power to change things in surprising ways. If you heat ice, it will turn into water, and then steam. The energy you add causes a spectacular change, turning a solid into a gas that looks nothing like the ice you started out with. Heat changes different substances in different ways. It can cook things, burn them, or melt them.

▲ EVAPORATING *Heating liquids makes part of them turn to gas and disappear, cooling the liquid left behind. This is evaporation. Sweating works by evaporation. Our bodies cool when sweat evaporates from our skin.*

▲ FREEZING *Most things shrink and go harder when you remove heat. Frozen food goes hard, while flowers cooled to low temperatures shatter like glass.*

▲ BURNING *Heat things with oxygen and you can start a dramatic chemical reaction called combustion (burning). Flames are made when the vapour (gas) given off by a burning object catches fire.*

▶ MELTING *Things that do not burn can melt instead. Melting means turning from a solid into a liquid. Water melts at 0°C (32°F), but bronze melts at 900°C (1,650°F).*

*Molten bronze glows orange or yellow when some of its heat energy turns into light*

*Insulated glove stops heat burning the person's hand*

*Sand is a good insulator that protects the surrounding surface from heat*

133

# Heat transfer

Coffee always goes cold and ice cream always melts. Hot things cool down and cool things warm up because heat is constantly on the move. Although heat transfer is sometimes a nuisance, it can help us to do many things. A refrigerator removes heat from food to keep it cool and fresh, while an oven heats the same food up to cook it. Heat transfer dries our clothes, indoors or out, and keeps our homes cosy in the depths of winter.

## WHY DOES HEAT MOVE?

Heat moves when things are at different temperatures. It always moves from hotter to colder things, never the other way. In winter, you are hotter than the air around you. Heat flows out from your body, and it is hard to stay warm. On summer days, heat flows into your body, and it is hard to keep cool.

**WOW!**

Nothing can get colder than –273°C (–459°F). Scientists call this temperature "absolute zero".

▲ BARBEQUE *Cooking food transfers heat into it. This kills germs, making food safe to eat. Cooking also makes food easier for us to digest.*

## HOW DOES HEAT MOVE?

Heat moves in three different ways. If you touch something hot, you can feel the heat moving into your body. This is heat conduction, and it happens whenever hot things come into contact with cooler ones. If you lean over a radiator, you can feel hot air rising. This is called convection, and it is the main way heat travels through liquids and gases. Heat can also beam straight through the air or empty space. This is called heat radiation.

▲ CONDUCTION
*When a saucepan sits on a stove, heat transfers from the stove to the pan by conduction.*

▲ CONVECTION
*Hot liquids rise above cooler ones. Heating this purple dye makes it float in the water.*

▲ RADIATION
*A hot lamp beams heat through the air in all directions, just as the Sun fires heat through space.*

## KEEPING WARM

The best way to keep warm is to stop heat escaping. This is called heat insulation. We wear layers of clothes in winter because they trap air between them. This reduces heat conduction, convection, and radiation from our bodies so we stay warmer. Animals do the same with thick layers of blubber and fur. The feathers on a penguin's coat trap air just like the layers of clothes we wear.

*Tightly screwed lid stops convection*

*Vacuum (empty gap) stops convection and conduction*

*Metal inner bottle stops radiation*

*Thin plastic stand stops conduction*

◄ SNUGGLING SCIENCE
*When a baby penguin cuddles up to its mother, heat conduction keeps it warm and cosy.*

## HOW VACUUM FLASKS WORK

A hot drink stays hotter much longer if you put it in a vacuum flask. Inside, there is a glass or metal bottle surrounded by a vacuum (empty space with no air inside) and plastic to stop heat getting out so quickly. Vacuum flasks also stop heat getting in, so you can use them to keep cold drinks cool as well.

## DRYING WITH HEAT

Laundry dries when the water inside it evaporates. We can dry washing outside using conduction, convection, and radiation. If we lay it on warm ground, it dries by conduction. If we hang it in the air, convection plays a part. When the Sun shines on washing, radiation warms the cloth and the water inside, helping the water to evaporate.

135

# Radioactivity and nuclear power

You can fit a million atoms on the head of a pin. Each one of these tiny bits of matter has the power to make a massive amount of energy. We can release this energy by smashing an atom apart or crashing two atoms together. Some atoms split all by themselves, giving off a kind of energy called radioactivity, which can be dangerous but very useful.

## WHAT IS RADIOACTIVITY?

Some atoms are unstable and break up to make more stable ones. When they split, they give off radioactivity. This can be chunks of atoms called alpha particles, flying neutrons called beta particles, or waves of energy known as gamma rays.

Lead

Aluminium

Hand

Alpha

◄ PENETRATING POWER *Alpha particles can be blocked by your hand. Beta particles are stopped by aluminium. It takes lead to stop gamma rays.*

Gamma

Beta

1. Hydrogen atoms fired in

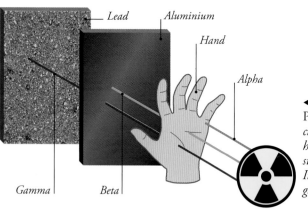

1. Neutron fired at high speed

2. Uranium atom splits apart when neutron hits it

3. Big atom splits into smaller atoms and releases more neutrons

## WHY DO ATOMS CONTAIN ENERGY?

Suppose you had all the bits to build an atom – a big pile of protons, neutrons, and electrons. You could squeeze them together but you would need to use massive force and energy to do so. If they stayed together, the energy you used would be locked inside the atom as potential energy. This is the energy that gets released when an atom smashes apart.

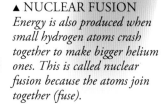

▲ NUCLEAR FUSION *Energy is also produced when small hydrogen atoms crash together to make bigger helium ones. This is called nuclear fusion because the atoms join together (fuse).*

◄ NUCLEAR FISSION *We can make energy by smashing big uranium atoms. Fission (splitting atoms) releases more neutrons that smash more atoms in a chain reaction.*

4. Neutrons split other atoms

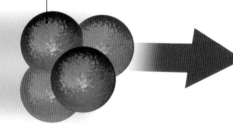

*4. Energy released*

*3. Spare neutrons fire off*

*5. Helium formed*

*2. Collision releases massive amount of energy*

## USEFUL RADIOACTIVITY

Radioactivity is very dangerous, but it can also be very helpful. In hospitals, radiation treatment is sometimes used to treat cancer. A beam of radiation is fired at tumours (harmful growths in the body) to kill them and stop them spreading. Food is sometimes treated in a similar way so it lasts longer in shops.

◀ FOOD PRESERVATION *Food can be blasted with radiation to kill germs so it lasts longer. Some people argue against this because radiation can also kill minerals and vitamins.*

ALBERT EINSTEIN

German-born scientist Albert Einstein discovered that a tiny amount of matter can change into a huge amount of energy. This is the basic idea that makes a nuclear power station work. It also explains why a nuclear bomb causes a massive explosion and a huge amount of damage.

# WOW!

Nuclear power stations make very dangerous waste. Some of it stays radioactive for more than a million years.

## HOW NUCLEAR POWER WORKS

Nuclear reactions release a lot of energy. We can use a nuclear reactor to turn this energy into useful electricity. Instead of burning fuel like an ordinary power station, a nuclear reactor uses energy from splitting atoms.

▶ NUCLEAR REACTOR *Heat is made in the dome-shaped nuclear reactor. This produces steam that turns the generator, making electricity that powers our homes.*

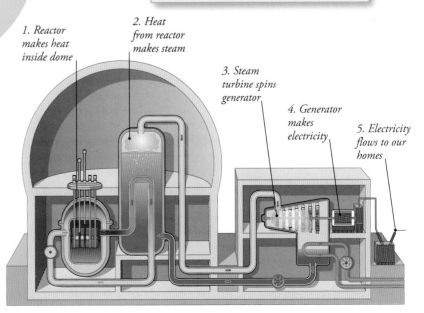

*1. Reactor makes heat inside dome*

*2. Heat from reactor makes steam*

*3. Steam turbine spins generator*

*4. Generator makes electricity*

*5. Electricity flows to our homes*

# Energy conversion

Energy lasts forever. It cannot be created or destroyed. But it can turn from one form into another. Every single burst of energy that beams down to Earth from the Sun has to go somewhere. All energy can change into other forms. It keeps changing, until it eventually becomes waste heat that we can no longer use.

## WOW!

Cars waste about 80 per cent of the energy they get from petrol. Only 20 per cent of the energy moves you down the road.

## ENERGY IN MOTION

Almost all our energy comes from the Sun. When sunlight hits the Earth, it makes plants grow, which feed animals, including humans. Packed full of food, animals have the energy to do all kinds of things. Even the simplest things an animal does (such as thinking or sleeping) can use up a lot of energy.

▲ SUNLIGHT *Almost all the energy we use on Earth comes from the Sun. A little bit of it also comes from heat made deep inside the Earth.*

▼ GROWING PLANTS *Plants are living "machines" that catch sunlight and store it. They do this using photosynthesis, a process which turns light into chemical energy.*

▶ FOOD *Wild animals can spend more than 12 hours a day eating. Humans do not eat as much because we can select the most energy-rich food, and cook it to give us maximum energy.*

▲ EATING *Your body digests food to extract energy and nutrients. Excess energy is stored as fat.*

## ENERGY IN, ENERGY OUT

When something loses energy, that energy has to go somewhere. When you feel hungry, it means you have used up the energy you took in from food. Some of it has kept you warm. More has been used by your brain for thinking. You have used some moving about. If you add up all the energy your body uses in different ways, it comes to the same amount as the energy in the food you eat to start with.

► RUNNING *Moving your muscles takes energy because you have to lift your heavy limbs against the force of gravity.*

▲ RESTING *Doing nothing takes energy because your body still has to think and keep warm. Even while resting, your body is busy on the inside.*

### JAMES JOULE

British scientist James Joule showed how energy can change from one form to another. In 1845, he proved that when something loses energy, another thing must gain the same amount of energy. This is called the "conservation of energy" and it is one of the most important scientific laws.

*The heart is the hottest part of a horse.*

▲ HEAT *Animals use a lot of energy keeping warm. This thermal photograph shows how much heat a horse's body loses. Red parts are hottest. Blue parts are coldest.*

▲ ACTIVITY *When an animal moves, it uses up energy that it has stored from food. Animals store energy as body fat, so they do not need to eat all the time.*

ENERGY

# Waves

When winds roar across the oceans, the water whips into towering waves that ripple around the world, until they crash on to the shore. Waves are how energy travels from place to place. Waves on the ocean are formed when energy from the wind is transferred to the water. Earthquakes also create giant waves, which can shake the surface of the land, and create towering tsunamis out at sea.

## WOW!

An ocean wave crashing on a beach in front of you may have travelled more than 15,000 km (9,000 miles) across the ocean.

## RIDING ON WAVES

Surfers can zoom across the ocean because waves are packed full of energy. The higher and faster the wave, the more energy it carries. A typical wave carries enough energy per metre of its width to power up to 1,000 large light bulbs. Waves break and release their energy when they hit the shallow water near the shore.

## ENERGY IN MOTION

Although ocean waves seem to move the water, the ocean does not go anywhere. In fact, the water just moves up and down. As it does so, energy passes sideways into the neighbouring water molecules. You can see the same effect with a friend by holding on to the ends of a piece of string. If one of you shakes the string, waves of energy will pass all along it.

*Water moves up and down*

*Energy moves sideways*

**How waves carry energy**

## EPIC EARTHQUAKES

Earthquakes are waves that shake the ground beneath our feet. Earth's land masses are built on giant rocky plates that grind slowly past one another. Sometimes they jolt up and down, causing a massive release of energy at cracks called faults. This energy races through Earth as a high-speed wave, producing an earthquake on the surface.

▲ FAST WAVES *The waves that carry earthquakes speed through the ground at around 25,000 km/h (15,000 mph), 25 times faster than a jumbo jet.*

ENERGY

2. Energy makes waves on ocean surface

3. Waves race over ocean at high speed

4. Near the shore, waves get steeper, up to 30 m (100 ft) or higher

1. Earthquake releases energy

## DANGEROUS WAVES

Earthquakes often strike under the ocean. A large quake jolts the water above it very violently. This sudden release of energy causes waves to race across the ocean's surface. The waves carry massive amounts of energy, and when they reach the shore, they can be higher than buildings. They continue on to land as a tsunami – a massive wave that can flatten everything in its path.

▲ HOW A TSUNAMI HAPPENS *Energy from an earthquake races through the ocean, producing giant waves that can destroy towns close to the shore.*

5. Buildings on the shoreline are in danger of being damaged by the waves

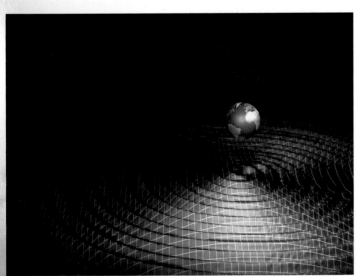

## WAVES IN SPACE

As well as physical waves, there are also waves of force and energy pulsing through the Universe. Some scientists believe that gravity makes waves that ripple through space and time, like waves on the ocean. Bigger objects, such as massive black holes, make bigger waves. As gravitational waves travel, they spread out and get weaker, which makes them very hard to detect.

◀ GRAVITATIONAL WAVES *Scientists are trying to find evidence that gravitational waves are racing through space. Although they have built sensitive wave detectors, no such waves have yet been found.*

# Sound

Close your eyes and listen to the world. You can recognize almost everything you hear because it has its own distinctive sound. Sounds are unique because they are all made in slightly different ways, but all sounds have one thing in common. They are created when something vibrates (shakes back and forth), sending waves of energy into our ears.

## GOOD VIBRATIONS

We make music with instruments that vibrate. Our voices are instruments too, singing when air vibrates inside our throats. The sound of a person talking, singing, or playing music is easy to listen to. Our ears pick up the sounds and our brains turn them into thoughts and feelings.

◀ MUSIC TO THE EARS *Bowing makes the strings of a violin vibrate, moving air inside the case that we hear as music.*

## BAD VIBRATIONS

Machines and engines have moving parts that grind and shake, making noise (sounds we do not want to listen to). Music is made from sound waves of varying pitch, arranged in a pleasant way. Noise is simply random sound, often at a single pitch.

◀ NOISE POLLUTION *Loud noise can damage your hearing. Workers usually wear earplugs to prevent sound waves from entering their ears.*

## LOUDNESS

Sounds travel as waves of vibration. The bigger the waves, the more energy they carry. When they arrive at your ears, louder sounds push harder against your eardrums. From rustling leaves to jet engines screaming past, our ears can "measure" an amazing range of quiet and loud sounds.

## WOW!

A space rocket taking off makes more noise than 10 million rock bands performing together.

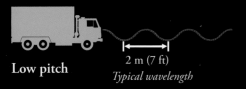

High pitch

11 cm (4 in)
*Typical wavelength*

Medium pitch

70 cm (30 in)
*Typical wavelength*

Low pitch

2 m (7 ft)
*Typical wavelength*

## PITCH

The sound something makes changes if it vibrates quicker or slower. When an object vibrates quickly, we hear high-pitched sounds. Low-pitched sounds come from things that vibrate more slowly. Most vibrating things make a mixture of many different sound waves. Everything makes a unique mixture – and that is why even two people's voices can sound very different.

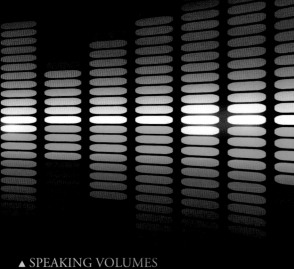

▲ SPEAKING VOLUMES
*The glowing light-emitting diode (LED) bars on a music player show the loudness of the sounds it plays. The more bars light up, the louder the sound.*

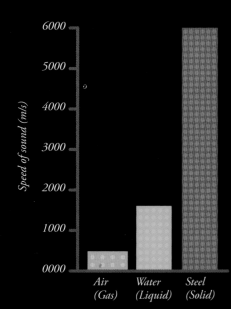

*Speed of sound (m/s)*

6000
5000
4000
3000
2000
1000
0000

Air (Gas)    Water (Liquid)    Steel (Solid)

## SPEED OF SOUND

Sound cannot exist in a vacuum – the vibrations that carry it need something to travel through. Surprisingly, sound travels much faster in solids than in liquids or gases.

◀ WHALE TALK *Whales communicate with one another by making low-pitched sounds in water. Their sounds can travel across entire oceans.*

## SONIC BOOM

As a plane flies, the noise of its engines speeds out in all directions. If the plane moves faster than the speed of sound, it starts to overtake the sound waves in front of it. A shock wave forms, spreading out in a cone behind the aircraft and creating a loud noise called a sonic boom.

# Music

When waves of energy vibrate inside your ears, you might feel like singing or dancing. We call this music. It is a special kind of sound that we enjoy listening to because it has the power to make us feel happy or sad. Music is made by instruments that shake the air so the sound rushes towards us. Most instruments make a range of sound frequencies, so they can play a musical tune.

WOW!

The oldest instrument ever discovered is a simple flute made from a vulture's wing bone. It was made 35,000 years ago.

## HOW INSTRUMENTS WORK

Musical instruments make sound by moving the air back and forth all around them. The faster they vibrate, the faster they shake the air and the higher the musical notes we hear. Most instruments are designed so they can vibrate at slightly different speeds, making many different notes. A guitar has six strings, but you can press them in different places to make dozens of different notes.

▶ BOTTLE ORGAN *You can build an organ by filling bottles with different amounts of water. When you blow into a bottle the air inside vibrates, making a musical note. The fuller bottles make higher notes while the ones with less water produce lower notes.*

ENERGY

146

## STARS OF THE ORCHESTRA

We can create an infinite number of melodies by combining different sounds from different instruments. Although each instrument makes sound waves, they all work slightly differently. Bigger instruments tend to make lower and louder notes than small ones. Instruments with more keys, strings, or holes can make a wider range of notes. Playing many instruments together in an orchestra makes even more interesting effects.

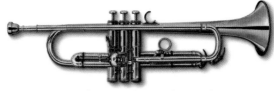

▲ BRASS *Air vibrates in the metal pipes of this trumpet to make sounds. Switches called valves are used to change the notes.*

▲ WOODWIND *Air vibrates inside a pipe when you blow across the end. The holes change the sound waves and the pitch of the notes they make.*

▲ PERCUSSION
*Banging a drum makes its tight elastic skin vibrate. The tighter or smaller the drum, the higher the pitch of the notes it makes.*

▶ STRING *A stretched string makes a note when you pluck it. Longer strings make lower notes than shorter ones.*

ENERGY

## ELECTRIC SOUNDS

It takes energy to produce sound, so making loud sounds for a long time is hard work. That is one reason why we have electric instruments. They use electricity to help us make loud sounds for long periods. Electric instruments also make very different sounds from traditional acoustic (non-electric) instruments.

▲ SYNTHESIZER *Electronic keyboards can copy the sounds of any orchestral instrument. They can also make futuristic sounds that no traditional instrument could ever play.*

## HOW DO ELECTRIC GUITARS WORK?

Ordinary acoustic guitars have strings that you pluck. When the strings move, they vibrate air inside the wooden case and this makes the sound. Electric guitars have metal strings with pickups (wire-wrapped magnets) underneath. When the strings vibrate, they make electric currents flow through the pickups. If a guitar is connected to an amplifier and loudspeaker, these currents are boosted in volume to make loud music.

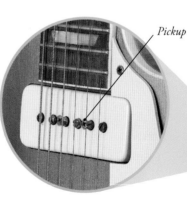

*Pickup*

▲ GUITAR PICKUPS *Each string on an electric guitar has a pickup underneath it, made from magnets and coils of wire. The pickups make tiny electric currents when the strings move above them.*

# LIGHT

## LIGHT DISPLAY
This train tunnel in Shanghai, China, is lined with colourful patterns, created by shining laser light beams on to the walls. These lights were photographed using a special effect to make them blur into streaks.

Light is energy radiating through space. This includes everything we see. From blazing sunlight to the tiny glow of a firefly, light brings life and colour to the things around us.

# Light and shadows

Light is the energy that shows us the world. People are daytime animals, and we live in a world of bright light and colour. Many animals prefer the dark and thrive when the Sun goes down. In between light and dark, there is a dusky world filled with shadows.

## LIGHT SOURCES
Light does not magically appear out of thin air – it has to be made using another kind of energy. A candle makes light by burning wax. The Sun makes its light when atoms smash together inside it, releasing huge amounts of nuclear energy.

## SEEING CLEARLY
We see things because light bounces off them and into our eyes. White things reflect most light, while dark things soak up the light and do not reflect much. In darkness, seeing things is very hard because little light reflects into our eyes.

*Source of light*

◄ OPAQUE OBJECTS
*Light cannot pass through a dull, opaque object. Some light gets absorbed and the rest reflects back into our eyes.*

*Light beam gets absorbed*

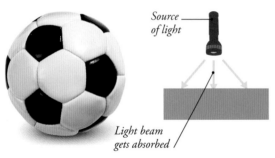

*Light is evenly reflected*

◄ SHINY OBJECTS
*No light is absorbed by a mirror. Instead, all the light reflects straight back again.*

*Most light passes through*

◄ TRANSPARENT OBJECTS *See-through things let most light pass through. But some light reflects back or we would not be able to see them.*

## SHADOWS
Light rays shoot in straight lines, and when something gets in their way, they cannot pass. A pattern of darkness forms behind the object, called a shadow. The dark middle of a shadow is the umbra. The lighter part around the edge is the penumbra.

# TIME FOR SHADOWS

People have been telling time using shadows for more than 5,000 years. As the Sun sweeps around the sky, it casts shadows on the ground. The shadow of a stick moves clockwise in a circle. At noon, it always points exactly north (in countries north of the Equator) or south (in countries south of the Equator).

▶ TELLING THE TIME
*Most sundials use a gnomon (tilted pointer). The gnomon and dial must be carefully lined up with a compass to tell the time correctly.*

*Penumbra or edge of shadow*

*Umbra or middle of shadow*

◀ SHADOW PUPPETS *Playing with shadows is one of the oldest games in the world. You can also create animal shadows by making shapes with your hands in front of a torch.*

## WOW!

During a solar eclipse, shadows sweep across the ground at up to 8,000 km/h (5,000 mph), 25 times faster than a racing car.

## SHADOWS IN SPACE

The Sun shines like a giant torch in space, but we can only see it if nothing gets in the way. Sometimes the Moon passes right in front of the Sun, casting a shadow over Earth. This phenomenon is called an eclipse.

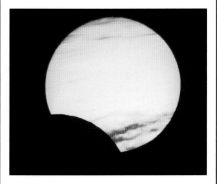

▲ SOLAR ECLIPSE BEGINS *As the Moon inches slowly across the Sun, it starts to block out the light.*

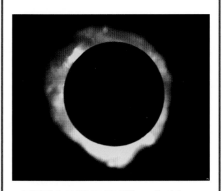

▲ TOTAL ECLIPSE *When the Moon is completely in front of the Sun, it makes a sparkle of light called the diamond ring.*

# Colour

Rainbows paint the sky with colour – but what exactly is colour? Things around us seem coloured because they reflect light into our eyes in different ways. The colours we see are made by light waves of different sizes (wavelengths and frequencies). Recognizing different colours helps our eyes and brains pick out objects to understand the world more clearly.

## LIGHT OF MANY COLOURS

White light is actually a mixture of all the other colours. We know this because white light splits into colours when you shine it through a wedge of glass called a prism (see pp.154–55). Raindrops and pieces of plastic also split white light into its colours in the same way.

▶ RAINBOW *When light from the Sun passes through raindrops, it bends by different amounts, splitting into different colours. Violet bends most so it is always on the inside of a rainbow.*

## WOW!

You might think a rainbow has only seven colours. In fact, there are an infinite number of colours between red and violet.

## CHANGING COLOURS

Daylight is made from white light. A tomato looks red because it reflects the red part of white light into our eyes and absorbs (soaks up) the other colours. If you change the colour of the light shining on it, you change the light that is reflected and absorbed. This makes the tomato look different and strange.

▶ WHITE LIGHT *The tomato absorbs green, blue, and all other colours and reflects red into our eyes, making it look red.*

▶ GREEN LIGHT *The tomato absorbs green light, so it turns black. The stalk reflects green light, so it still looks green.*

▶ RED LIGHT *The tomato reflects red light and so does the white background. The green stalk absorbs red light, so appears black.*

## SOAPY SWIRLS

Have you ever noticed colourful patterns swimming in soap bubbles? Light rays reflect off both the top and bottom of the thin soap film (wrapped around the bubble). These separate rays meet and merge together. This makes colours that change according to how thick the soap is. As the thickness of the film is constantly changing, the colours slowly change too.

▲ BLOWING BUBBLES *The film gets thinner as you blow on it, which makes the colours change from purple (where it is thickest) to yellow and clear (where it is thinnest).*

## HEAT MAKES COLOUR

Hot things glow red, yellow, or white hot because heat makes the atoms inside give off light. The hotter something is, the more the atoms are "excited" and the more light they give off. When something is white hot, it has so much energy that its atoms are giving off all the different colours of light at once.

◀ HOT STUFF *Hot things give off energy you can see (light) as well as energy you can feel (heat).*

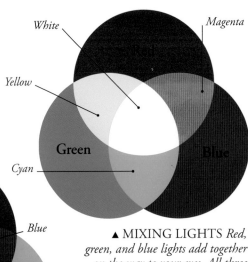

▲ MIXING LIGHTS *Red, green, and blue lights add together on the way to your eyes. All three together produce white light.*

## MIXING LIGHTS, MIXING INKS

Colours can change when light rays mix. If you shine red, green, and blue torches together, their light combines to make white light. If you mix red, green, and blue ink on a page, they absorb all the light that falls on them, so the page looks black.

◀ MIXING INKS *Different coloured inks absorb different wavelengths of light. By mixing them together, we create ink that absorbs all colours, so looks black.*

## PIGMENTS AND PAINTS

Pigments are the chemicals that give paints their vibrant colours, and most are made from salts (compounds) of metals. Titanium dioxide, found in sand, is used to make a brilliant white colour, while iron oxide makes paints that are red or brown. Green paint is made from chromium oxide.

## SPLITTING LIGHT

Our eyes see different wavelengths of light as different colours. White light is actually lots of wavelengths mixed together. When light passes through a prism it changes direction (refracts). Different wavelengths refract by different amounts, splitting white light into rainbow colours.

# Lasers

Lasers are light beams that fire out from a tube packed with energized atoms. Often used in light shows, lasers are also the power behind CD and DVD players, computer printers, and holograms, which make ghostly images dance before your eyes. Powerful lasers can leap miles into the sky and even slice through metal.

## HOW DOES A LASER WORK?

Lasers are concentrated beams of light. A laser works by flashing light into a special tube. As the atoms inside the tube soak up the light energy, they start to give off light themselves and cause other nearby atoms to give off light too. The rays of light all line up inside the tube and shine out in a single concentrated beam.

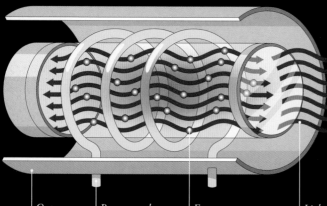

Outer case

Power supply flashes energy into tube

Energy causes atoms in tube to emit light

Light comes out focused into laser

## READING AND WRITING WITH LIGHT

When you learned to read, you might have moved your fingers under the words to keep your place. Laser beams read music and movies from CDs and DVDs in a similar way. Inside the player, the CD or DVD spins at high speed. Just like a tiny finger, a laser beam scans across the disc and reads information from bumps on the surface.

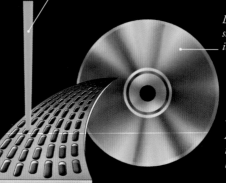

Pinpoint laser beam reflects off bumps on CD's surface

Disc is made of shiny aluminium so it reflects laser light

◄ PINPOINT LIGHT
*A CD stores over an hour of music in tiny bumps. There are up to 5 billion bumps printed in the shiny surface.*

## LEAPING LASERS

Lasers make more powerful light than torches, so they shine much further. In a torch beam, the light waves are jumbled up, so they tend to cancel one another out. They quickly soak into the air and do not go far. In a laser beam, the light waves all line up, so they are brighter and can travel further.

▶ LIFELIKE
*Colourful holograms bring science to life in a museum in Paris.*

## HOLOGRAMS

Photographs are flat images that always look the same. Holograms are like 3D photos that look different from every angle. They are made by bouncing lasers off objects to capture more information, which is displayed inside plastic or glass. Holograms have been used to make life-size 3D pictures of actors that can stand on stage. They are also used to stop people copying credit cards and banknotes.

◀ LASER SHOW *The light from lasers is so concentrated it almost looks like a solid object.*

## WOW!

The world's biggest laser is made from 192 laser beams and makes as much power as 10 billion light bulbs.

# Reflection and mirrors

Mirrors play throw and catch with light. They catch light from in front of them and throw it back the way it came. Mirrors are made from thin metal sheets inside glass, but many other surfaces reflect light as well. A smooth lake mirrors the sky above it, and you can often see your face in shiny shoes or a polished spoon.

## MIRROR SHAPES

Light rays stream in straight lines, so a flat and smooth (plane) mirror reflects things much as they are. The reflected picture looks like the original because the incoming rays bounce back in parallel lines. However, a mirror that curves inwards (concave) makes things look bigger, while one curved outwards (convex) can make them look smaller.

▲ PLANE MIRROR
*Light rays bounce back in parallel lines, so your reflection looks like you do. It seems reversed, left to right, so writing looks like gibberish.*

▲ CONCAVE MIRROR
*These are bent ("caved") inward and sometimes flip things upside down. They make things look bigger, so they are often used for shaving and bathroom mirrors.*

▲ CONVEX MIRROR
*These are bent outwards and make things look smaller and further away. They are used in car wing mirrors so drivers can see a wider view.*

**WOW!**

The giant mirror in the Hubble Space Telescope took five years to grind and polish.

## HEAT MIRRORS

Mirrors reflect heat as well as light. Hot objects give off infrared radiation, which is like invisible, hot light. When infrared hits a mirror, it reflects straight back again. You can test this effect for yourself by wrapping some silver foil around your arm. It feels warm because your body heat is being reflected back.

▲ HEAT BLANKET *Clothes keep you warm by trapping air that insulates your body. This metal blanket keeps you warm by reflecting body heat.*

*Each hexagon in the mirror weighs about 20 kg (46 lb).*

## ANIMAL MAGIC

Cats are like walking mirrors. They can see at night because they have special reflecting surfaces (mini mirrors) behind their eyes. These catch incoming light and bounce it back out through their eyes again. The light passes through their eyes twice, and this gives them double the chance to see dim objects. That is why a cat's eyes shine at night or in dim light.

## MEGA MIRRORS

Some telescopes need giant mirrors, but if a mirror becomes too big, it bends and buckles. To get around this, the biggest and best space telescopes use mirrors split into dozens of honeycomb-like segments. Bolted on to a framework, very close to one another, these work together like a single giant mirror.

◄ SEEING STARS *The James Webb Telescope, which will replace NASA's Hubble Space Telescope, is being made from 18 giant hexagonal mirrors.*

159

# Refraction and lenses

Lenses play tricks with light, bending it so it seems to come from a different place. The science behind this "magic" is called refraction, and it can make things look bigger than they really are. Lenses are the power behind microscopes and telescopes, which bring the world nearer. Working the opposite way, powerful lighthouse lenses can sweep beams of light far into the night.

## REFRACTION IN ACTION

Straight things often look wonky when they stand in water – and refraction is to blame. When light moves from water to air, it changes speed and bends in a new direction. This can make something underwater appear to be in a different place.

◄ BENDY PENS *Light from the lower halves of these pens is shifted by refraction, so they seem to be in the wrong place.*

## HOW A LENS WORKS

A drop of water can work just like a lens. Place some clear plastic on a newspaper, drop some water on top, and you will find it slightly magnifies the words underneath. A magnifying glass works the same way but makes the words much bigger, because it uses thicker plastic that can bend the light more.

WOW!

The world's biggest camera lens is 1.9 m (6 ft) long, weighs 100 kg (220 lb), and can magnify things more than 1,000 times.

◄ WATER LENS
*The water drop on this leaf bends light rays coming out. Like a lens, it makes the veins of the leaf look bigger.*

## SEEING THE LIGHT

If you want to throw a beam of light far out to sea, you need a powerful lens. A normal lens this big would be too heavy, so lighthouses use Fresnel lenses instead. With a surface made of sharp steps, they are much thinner and lighter. Each step bends the light slightly more, making a powerful, parallel beam. Fresnel lenses are also used in car headlamps.

▼ FRESNEL LENS *This lighthouse has a bright gas lamp behind a huge Fresnel lens. The lens spins on an electric turntable that sweeps its beam across the sea.*

## THICK AND THIN LENSES

Different-shaped lenses bend light in different ways. Lenses that are thick in the middle and thin at the edges make light rays converge (come together) at a point. Magnifying glasses, binoculars, and glasses for short-sighted people use convex (converging) lenses. Lenses of the opposite shape – thin in the middle and thick at the edges – make light rays diverge (spread out). Film projectors use concave (diverging) lenses.

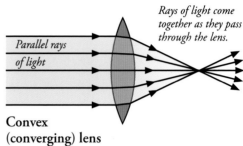

*Rays of light come together as they pass through the lens.*

*Parallel rays of light*

**Convex (converging) lens**

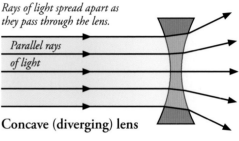

*Rays of light spread apart as they pass through the lens.*

*Parallel rays of light*

**Concave (diverging) lens**

## WHAT MAKES A MIRAGE?

Stare at a road on a hot day and you might see water in the distance. Move closer and there is no water there – it is a mirage. What you can see is light from the sky bent down into your eyes by layers of hot and cold air. The air makes light bend like a lens.

▲ JUST DESERT *The wobbling pattern in the distance is a mirage – a reflection of the sky in the hot desert air.*

# Telescopes and microscopes

Turn your eyes to the sky and you will see stars so far away that their light has taken millions of years to reach you. Telescopes make these pinpoints of light bigger, showing us a vast universe beyond Earth that we have barely explored. Back on our own world, there is another, tiny universe too small to see with the naked eye. Microscopes help us zoom in on this secret world.

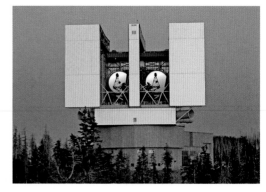

## GIANT TELESCOPES

The further you want to see, the bigger the telescope you need. Linking two telescopes makes a combined telescope that is more than twice as powerful. The Large Binocular Telescope in Arizona, United States (above) does this with two side-by-side mirror telescopes, each 8.4 m (28 ft) wide.

## HOW DOES A TELESCOPE WORK?

A telescope is a long tube with lenses at each end. The lenses catch light rays coming from a faraway object, and bend them so they appear to come from much closer. This makes the object look bigger. Large, powerful lenses can bend light more and make bigger images. Unfortunately, large lenses are heavy and expensive, and they can distort images. That is why most large telescopes use mirrors to make their images instead.

**WOW!**

The world's largest radio telescope, at Arecibo in Puerto Rico, is 305 m (1,000 ft) across – three times the length of a soccer pitch.

## SEEING WITHOUT LIGHT

Distant stars do not just beam out light we can see. They also give off other kinds of energy (electromagnetic radiation) such as radio waves, X-rays, and microwaves. If we fit telescopes with sensors that can pick up these kinds of energy, we can build up much more detailed pictures of the stars and planets than we get from ordinary telescopes.

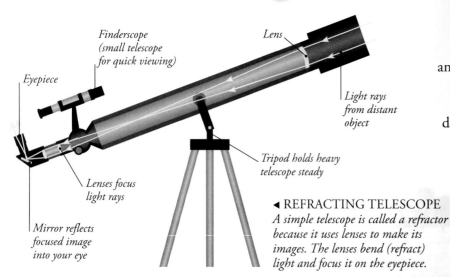

Finderscope (small telescope for quick viewing)

Lens

Eyepiece

Light rays from distant object

Lenses focus light rays

Tripod holds heavy telescope steady

Mirror reflects focused image into your eye

◄ REFRACTING TELESCOPE
*A simple telescope is called a refractor because it uses lenses to make its images. The lenses bend (refract) light and focus it on the eyepiece.*

## HOW DOES A MICROSCOPE WORK?

Like a telescope, a microscope is a tube with mirrors and lenses inside. It catches the light rays given off by tiny things and spreads them apart, so they look bigger and our eyes can see them more clearly. A microscope works the same way as a magnifying glass, but uses extra lenses to make a more magnified image.

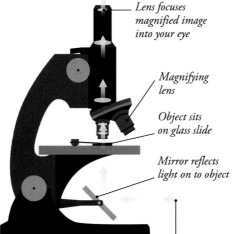

Lens focuses magnified image into your eye

Magnifying lens

Object sits on glass slide

Mirror reflects light on to object

Light shines in from lamp

## GIANT MICROSCOPES

Just as there are huge telescopes, so there are huge microscopes. The Diamond Light Source near Oxford, UK, is one of the world's biggest. It whirls electrons around a giant ring to make powerful beams of radiation. These are finer than light waves, so they can see things in more detail.

▲ DIAMOND SYNCHROTRON
*Electrons race around a ring of magnets inside this doughnut-shaped building at just under the speed of light. That makes them fire out beams of light 10 billion times brighter than the Sun.*

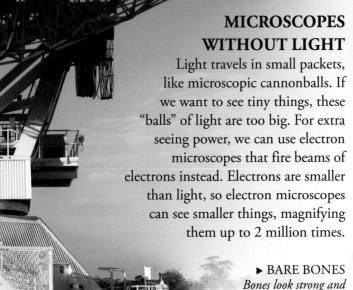

*This huge radio telescope, called the Telescope Compact Array, is located in New South Wales, Australia. It has six dish-shaped antennae, such as this one, each 22 m (72 ft) in diameter.*

## MICROSCOPES WITHOUT LIGHT

Light travels in small packets, like microscopic cannonballs. If we want to see tiny things, these "balls" of light are too big. For extra seeing power, we can use electron microscopes that fire beams of electrons instead. Electrons are smaller than light, so electron microscopes can see smaller things, magnifying them up to 2 million times.

▶ BARE BONES
*Bones look strong and solid on the outside, but an electron microscope can zoom in to show the spongy structure inside.*

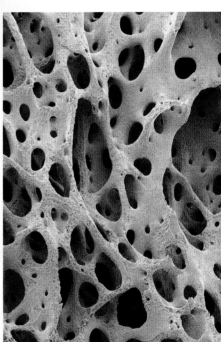

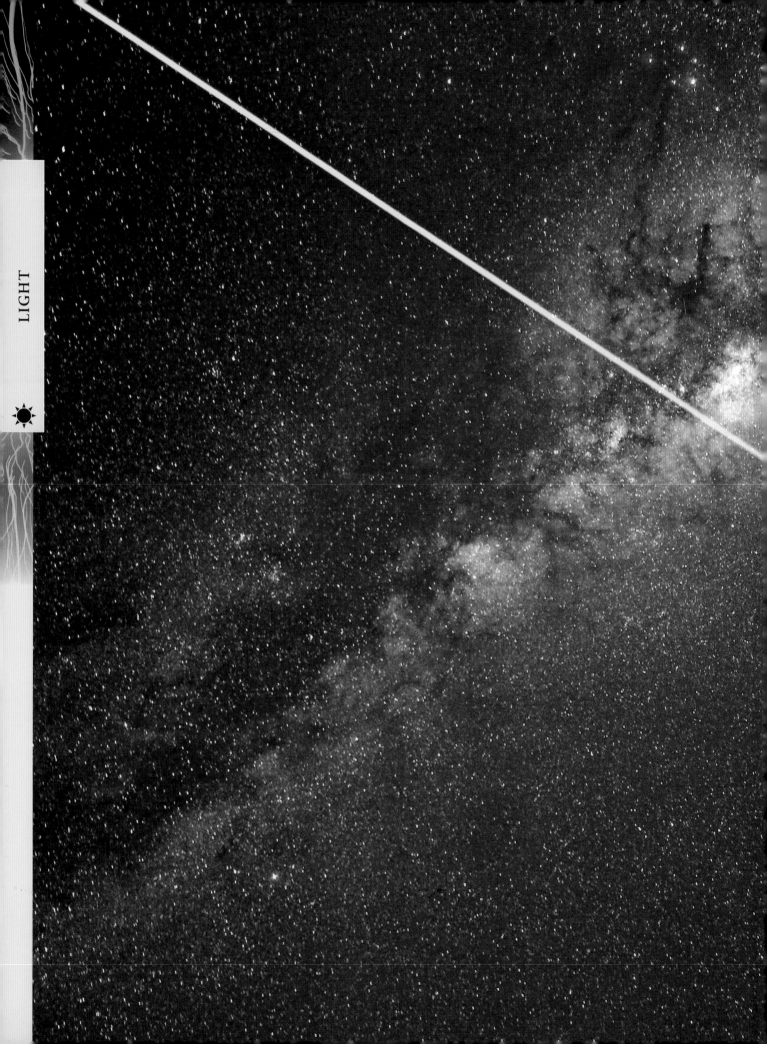

**LASER-GUIDED TELESCOPE**
This is one of the four telescopes that
make up the Very Large Telescope –
one of the world's most powerful
star-gazers. It uses laser beams to help
it aim at distant stars and planets. It
is built high up in the Atacama Desert
in Chile, where the air is thin and
clear, and the stars are easier to see.

# Cameras

We see things because light bounces off them and into our eyes, where our brains recognize them. Cameras are like artificial eyes. They can record patterns of light as images and store them for us to look at later. They use lenses to take in light from a wide area and focus it on a small recorder.

## HOW DOES A DIGITAL CAMERA WORK?

Cameras record light reflecting off things. They do this using lenses and mirrors, which focus the light on to an image sensor – an electronic chip made of light-detecting cells. When light falls on the cells, they convert it into a pattern of numbers that the camera stores.

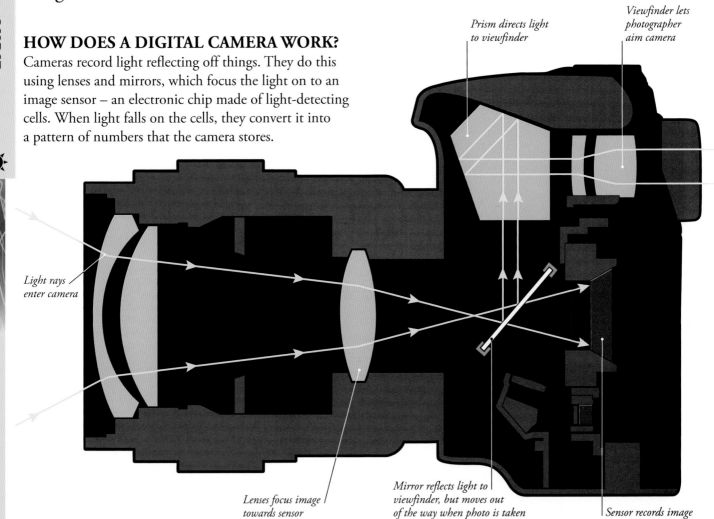

*Prism directs light to viewfinder*

*Viewfinder lets photographer aim camera*

*Light rays enter camera*

*Lenses focus image towards sensor*

*Mirror reflects light to viewfinder, but moves out of the way when photo is taken*

*Sensor records image*

## PINHOLE CAMERAS

Modern cameras are based on an older invention called the pinhole camera. It is a sealed box with a tiny hole on one side. As light streams through, it makes an upside-down image on the back wall of the box. You can even project an image on the wall of a room by making a tiny hole in the blinds. Pinhole cameras were invented in Ancient China, but the first person to understand how they worked was Islamic scientist Alhazen, around 1,000 years ago.

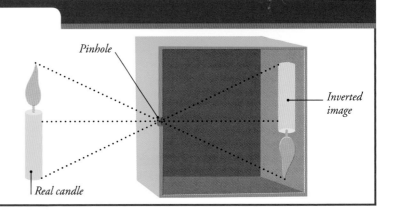

*Pinhole*

*Inverted image*

*Real candle*

## EARLY PHOTOGRAPHS

Photography was born in 1727 when a German scientist called Johann Schulze found that chemicals containing silver looked different after light shone on them. A century later, photographers were able to capture the first images of the world around them.

▶ CAUGHT ON FILM
*Frenchman Louis Daguerre made the first practical photographs in 1831. He used silver-coated plates to make detailed pictures called daguerreotypes. This one is thought to be the first photograph of a living person.*

## EXTREME PHOTOGRAPHY

Ordinary snapshots capture the world as we see it, but our eyes cannot see everything. They are not fast enough to catch things that happen very quickly, or sharp enough to notice very tiny changes. Special photography techniques can take astonishing pictures that our eyes alone could never see.

**WOW!**
Some cameras can take a photo in just 1/8,000th of a second – much faster than the eye can blink.

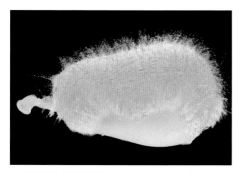

▲ SLOW-MOTION *Cameras can slow things that happen too fast for our eyes to see. This photo shows what really happens when a balloon full of water bursts.*

▲ MACRO *Powerful lenses can zoom in on the details of microscopic things. Macro lenses make large, detailed close-ups of very small objects, such as these seeds.*

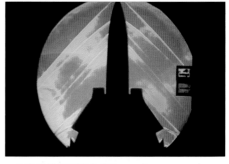

▲ SCHLIEREN *Schlieren photos use coloured lines to reveal invisible air moving around things. They are used to design more efficient shapes for aeroplanes and spacecraft.*

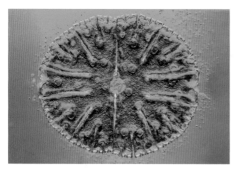

▲ MICROGRAPH *A photograph taken under a microscope is called a micrograph. With an electron microscope, we can make lifelike 3D micrographs, such as this photo of an algae.*

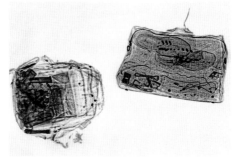

▲ X-RAY *X-rays pass straight through soft materials such as fabric and skin, so they can show up the ghostly details inside things, like the contents of these bags.*

▲ LONG EXPOSURE *Photos taken over a long time show moving lights as lines. These circles are trails made by stars circling the sky.*

# Moving pictures

What is it like to be a pirate out at sea, to live in the future, or to rocket into space? Movies use tricks of light to show us these exciting stories. A movie is made from many still photographs that flash in front of our eyes very quickly. Our brains merge them together to make realistic moving pictures.

## MOVIE-MAKER

An ordinary camera captures one picture at a time. A movie camera takes many pictures, one after another. Old cameras store their pictures on a reel of plastic film. Modern cameras save pictures on memory chips like the ones used in computers.

*Screen plays back recorded images very quickly, so they look as though they are moving*

▶ CAMCORDER *This portable movie-making machine stores pictures on a memory card. You can watch your movie on the screen and copy it to your computer as well.*

*Camera takes 60 pictures every second, and records sound*

**1. Camera lens captures light from object**

**2. Microchip in camera turns light to numbers**

**3. Memory card in camera stores numbers**

### STILL MOVIES

Movies were invented when people found out how to take quick photos, one after another. Often, the photos were printed on a strip. British photographer Eadweard Muybridge made many film strips, including these galloping horses taken in 1878.

## HANDS-ON MOVIES

"Stop-motion" is a way of making movies using small clay models. The models are photographed, then bent into a new position, and photographed again. Each second of animation needs 12 photographs or "frames".

▶ CLAY CHARACTERS *These characters from the movie* A Close Shave *(1995) are made of modelling clay, but seem to move on screen.*

## COMPUTER ANIMATION

How can you make a car laugh or a panda dance? You cannot film things that do not exist, but you can draw them on a computer screen and use computer software to turn them into moving pictures. This is called computer-generated imagery (CGI).

◄ DIGITAL ACTORS *These characters from* Kung Fu Panda *(2008) were created and animated on computers.*

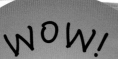
**WOW!**
A full-length "stop-motion" movie contains up to a million separately photographed frames.

## FLYING ON THE SPOT

Computer screens create moving pictures in the same way as movies, by flashing up lots of still images very quickly. When pilots learn to fly aeroplanes, they sit in fake cockpits with computer screens all round the walls. As the pilot moves the controls, the view changes as though the aeroplane were flying.

▼ HAPPY LANDING *A pilot practises landing an Airbus A330 plane on a flight simulator in Berlin, Germany. Every detail of the cockpit is copied exactly from a real aeroplane.*

169

# Communicating with light

It takes about seven hours to fly from London to New York City. A beam of light can make the same journey more than 50,000 times over in less than a second. Nothing travels faster than light, which is why this sparkling form of energy is the perfect way to send messages.

### INSTANT LIGHT

You will see the flash of lightning seconds before you hear the crash of thunder. The sound and the light set off together, but the light travels almost a million times faster than the sound. Light is almost instant. That is why we use light for most forms of communication.

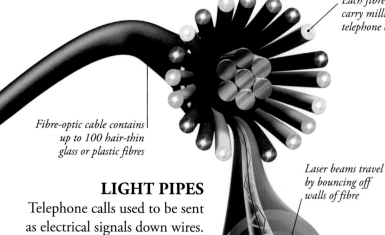

*Each fibre can carry millions of telephone calls*

*Fibre-optic cable contains up to 100 hair-thin glass or plastic fibres*

*Laser beams travel by bouncing off walls of fibre*

### LIGHT PIPES

Telephone calls used to be sent as electrical signals down wires. Electricity takes time to travel, and there was often a delay sending calls between countries. Now calls are sent by flashing lasers down fibre-optic cables (thin glass pipes).

## JOURNEY INTO YOUR BODY

It can be hard for a doctor to see why someone is ill if the problem is hidden deep inside their body. Endoscopes solve this problem. They are cameras with long, bendy fibre-optic cables attached. The cable is poked inside the patient, and light shines down it. The light reflects back up a second cable, sending back a clear picture that the doctor can see.

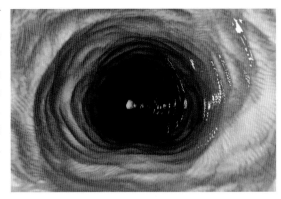

▲ TUBE TRAVEL *Your oesophagus (the tube from your throat to stomach) looks like this through an endoscope.*

## SIGNALLING WITH LIGHT

Light is the oldest way of communicating. Before telephones or the Internet, people sent messages between towns by lighting huge fires (beacons). Before radio, armies could send signals by reflecting the Sun off mirrors. Although this might seem slow, these signals race to your eyes at the speed of light.

◄ EMERGENCY LIGHT *This signalling lamp uses dozens of small light-emitting diodes (LEDs) to make a single powerful beam.*

WOW!

A beam of light can leap the 150 million km (90 million miles) from the Sun to Earth in just over 8 minutes.

## GLOWING CREATURES

Sound might be important to humans, but for the rest of the natural world light and colour are much more important forms of communication. Some animals can create light inside their bodies to scare away predators. Fireflies glow in the dark to attract a mate. Squid can change colour to show their mood.

▲ FLASHING FISH *Jellyfish are bioluminescent. They make flashes of blue light inside their bodies to scare off predators.*

# ELECTRICITY
# AND MAGNETISM

**MAGNIFIED MICROCHIP**
Computer chips use electricity to
record information and perform
mathematical calculations. They do
this by using millions of tiny switches,
which are so small they can only be
seen under a microscope.

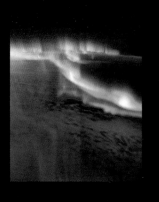

Electricity floods your home with light, while magnetism can pick up cars or stick things to your refrigerator. Working together, they can create amazing tools, from motors to supercomputers.

# Electricity

Everything electrical is powered by tiny electrons, the mini particles that whirl around atoms. In ancient times, no one really understood electricity or how it could be used. Now we know electricity is created when electrons gather together or zap from place to place. Electricity can help us do all kinds of things, from powering trains to catching criminals.

## WHAT IS ELECTRICITY?

Electricity is energy caused by electrons. When electrons gather in one place, they make static electricity (the kind that crackles in your jumper when you take it off). When electrons move about, they make current electricity. This is what powers things like vacuum cleaners and torches.

▶ LEAPING LIGHTNING *During a storm, clouds rub through the sky, and build up static electricity. When the clouds have too much charge to hold on to, they release it through lightning bolts (current electricity), which crash down to the earth.*

## POSITIVE–NEGATIVE

Scientists have studied electricity for hundreds of years. They soon learned that you could give some objects an electric charge by rubbing them. If you charge two of the same object, such as a plastic ruler, they push away from each other. If you charge two different objects, such as a plastic ruler and a glass ruler, they pull towards each other. American scientist Benjamin Franklin called this positive and negative electrical charge, and we still use those labels today.

◀ BENJAMIN FRANKLIN *This American scientist studied electricity in the 18th century and made several important discoveries.*

*Lightning branches out like a tree as the electricity in a cloud tries to find the quickest way down to earth.*

## FLOATING TRAINS

Scientists have discovered that some substances conduct electricity perfectly. None of the electrical energy is lost passing through them. These are called superconductors, and they can be used to create incredibly strong magnets. The world's fastest trains use these powerful magnets to float above the surface of the track, so there is no friction to slow them down.

## CONDUCTORS AND INSULATORS

We use copper to make electrical wires because metals such as copper carry electricity very well. They are good conductors, which means electricity can go through them easily. Other materials, such as rubber and plastic, stop electricity going through. These are called insulators. Electrical wires are made of good conductors, but often have insulators wrapped around them to protect us from electric shocks. Semiconductors are substances that can conduct or block electricity at different times, making them useful in electronics.

*Rubber*

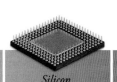
*Wood*

*Silicon*

*Water*

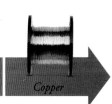
*Copper*

**Insulator**  **Semiconductor**  **Conductor**

*Dusting powder with a very fine brush reveals hidden fingerprints without destroying them.*

## ELECTRIC FINGERPRINTS

When criminals leave fingerprints, they also leave electricity. Fingerprints contain proteins that have a positive electric charge. Detectives can find prints by sprinkling them with gold dust, which has a negative electric charge. The opposite charges attract, the gold sticks to the proteins, and the hidden fingerprints become visible.

**WOW!**

There is enough electric power in a lightning bolt to boil water for 50,000 cups of coffee.

▶ DUSTING FOR PRINTS *Forensic scientists (people who apply scientific knowledge to solve crimes) use the power of static electricity to find fingerprints hidden on everyday objects.*

175

# Circuits and current

To make electricity do useful things, we have to push it around a loop. This is a circuit – the secret power behind everything electrical in your home. The looping electricity is called a current and the force that pushes it is a voltage. Sometimes the voltage is made by a battery, although in our homes it also comes through sockets carrying electricity from power stations.

**ELECTRICITY AND MAGNETISM**

### GO WITH THE FLOW

Electric things work when current flows through their circuits. A current is really a flow of electrons, zipping down a wire like water whooshing down a pipe. The main purpose of a circuit is to transport energy. The energy flows from a battery (or other power source), around the wire, to a device such as a lightbulb or electric motor.

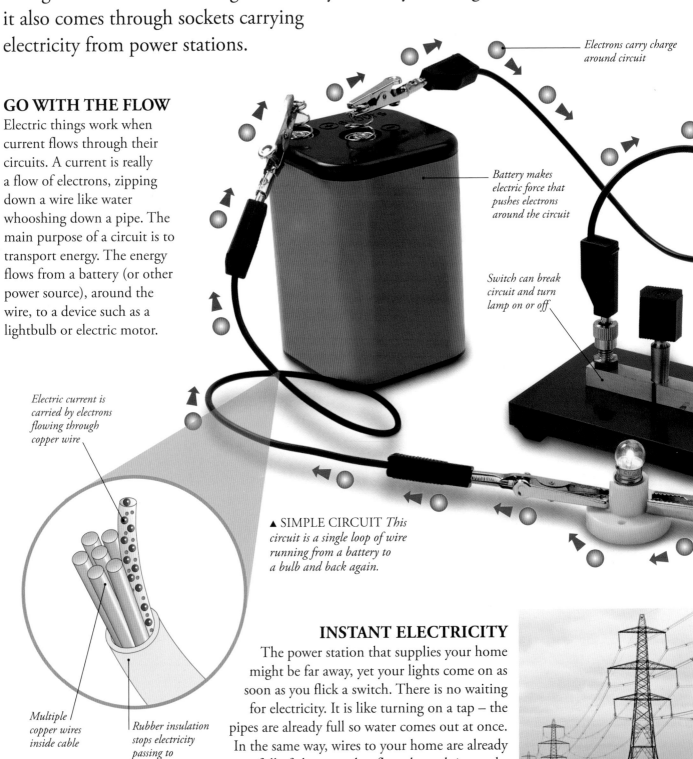

*Electrons carry charge around circuit*

*Battery makes electric force that pushes electrons around the circuit*

*Switch can break circuit and turn lamp on or off*

*Electric current is carried by electrons flowing through copper wire*

▲ SIMPLE CIRCUIT *This circuit is a single loop of wire running from a battery to a bulb and back again.*

*Multiple copper wires inside cable*

*Rubber insulation stops electricity passing to surrounding objects*

### INSTANT ELECTRICITY

The power station that supplies your home might be far away, yet your lights come on as soon as you flick a switch. There is no waiting for electricity. It is like turning on a tap – the pipes are already full so water comes out at once. In the same way, wires to your home are already full of electrons that flow through instantly.

176

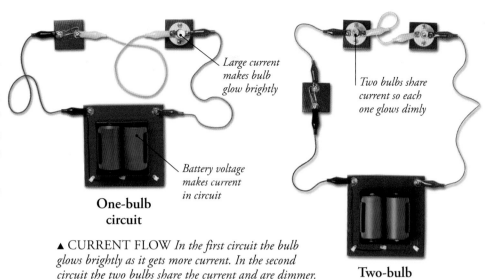

# VOLTAGE AND CURRENT

Voltage pushes current around a circuit. The bigger the voltage, the bigger the current. Electric things that work harder need higher voltages and currents. An electric cooker needs a much bigger voltage and current than a mobile phone. Similarly, a phone needs more voltage and current than a torch.

*Large current makes bulb glow brightly*

*Two bulbs share current so each one glows dimly*

*Battery voltage makes current in circuit*

**One-bulb circuit**

**Two-bulb circuit**

▲ CURRENT FLOW *In the first circuit the bulb glows brightly as it gets more current. In the second circuit the two bulbs share the current and are dimmer.*

**WOW!**

Every second that a small torch bulb is lit, a million trillion electrons stream through it.

*Open switch stops current flowing*

*Battery*

## SWITCHES

You do not want your kettle boiling all day long or your television blaring out at night. Switches help us turn circuits on and off when we need to. They work just like bridges, allowing current across when they are closed and stopping it from flowing when they are open.

*Bulb glows when circuit is closed but not when it is open*

*Circuit*

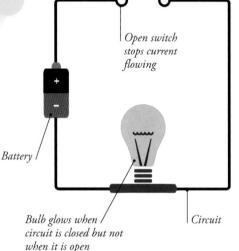

## ELECTRIC EEL

Electric eels are like batteries that can swim – they can make instant electric current, though only for a moment. A typical eel can make up to 600 volts, which is much higher than the voltage used in our homes. The sudden crack of an electric shock helps an eel fight predators and stun its prey, though they rarely harm humans.

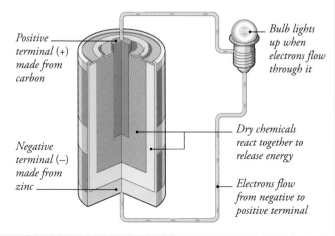

### BATTERIES

Batteries are like miniature power stations you can carry around. Inside, they are packed full of chemicals. When you connect batteries into a circuit, the chemicals react together and make electricity. A battery will keep making electricity for as long as the chemicals last.

*Positive terminal (+) made from carbon*

*Bulb lights up when electrons flow through it*

*Negative terminal (–) made from zinc*

*Dry chemicals react together to release energy*

*Electrons flow from negative to positive terminal*

# Static electricity

Whenever you see stormy lightning leaping to the Earth, you are watching a sudden zap of static electricity. This is the kind of electricity that happens when many electrons gather in one place. No one has yet managed to work out how to capture lightning, but static electricity has plenty of other uses. It can power photocopiers and printers, and scrub black smoke from chimneys.

*1. As you rub on a car seat, electrons flow from the seat to your body.*

*2. The electrons cling to your clothes and body, even when you get out of the car.*

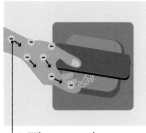

*3. When you touch the metallic handle, the electrons can flow into it, giving you a shock.*

## SHOCKING STUFF

Have you ever had a tiny shock when you touched a door handle? It happens because your body builds up static as it rubs against things. The static stays on you until you touch something metal. Then it moves from your body, through the metal, to the Earth – giving you a shock.

## HAIR RAISING

Electrons are little particles of negative electric charge. Some substances can take on more electrons by rubbing against things. Taking on more electrons makes them negatively charged, while the object that loses electrons becomes positively charged. Charged objects can create forces. Two things with the same charge will push away from each other, but two things with opposite charges will pull towards each other.

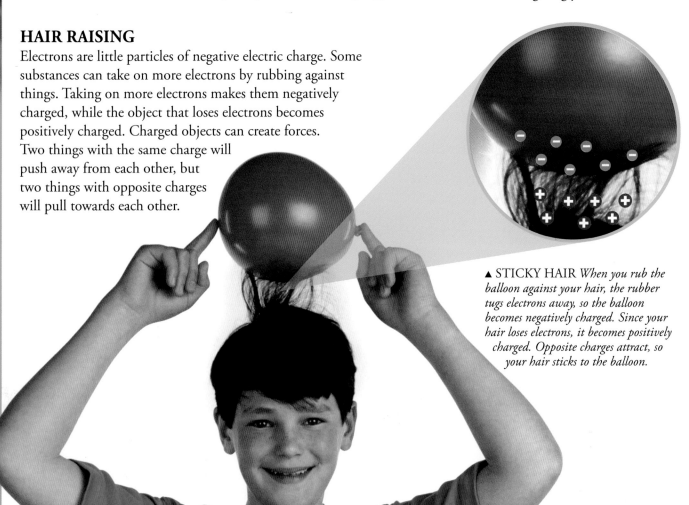

▲ STICKY HAIR *When you rub the balloon against your hair, the rubber tugs electrons away, so the balloon becomes negatively charged. Since your hair loses electrons, it becomes positively charged. Opposite charges attract, so your hair sticks to the balloon.*

# CAN YOU SEE ELECTRICITY?

Magnets make invisible patterns around them, called magnetic fields. When static electricity builds up, it makes a similar pattern called an electric field. If you rub a balloon to give it a static charge, an electric field forms around it. If you put tiny pieces of paper nearby, they are inside the balloon's electric field, so they get "sucked" towards it by force and stick to it.

▶ BLOOD TEST *Electric fields can be used to test people's blood for diseases. Doctors put the blood on a special microscope slide and dip an electric wire into it. The wire creates an electric field that draws patterns in the blood. The colours show if the patient is healthy or ill.*

*Electric field separates colourful chemicals out from the blood* —

*Clean white smoke comes out of chimney*

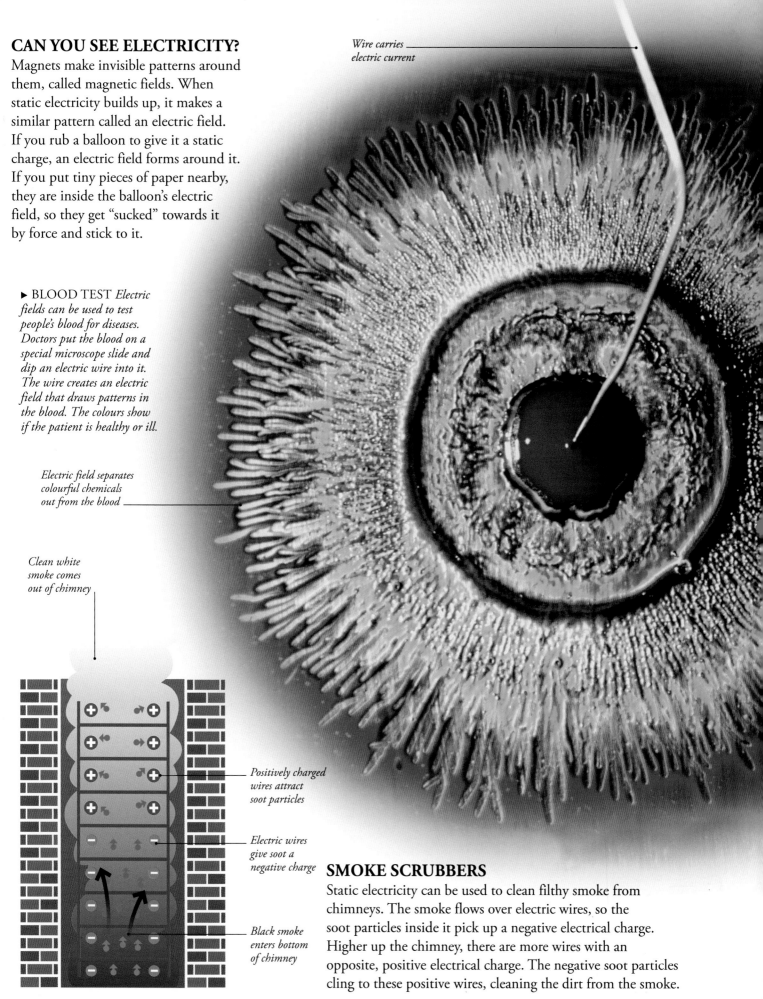

*Wire carries electric current*

*Positively charged wires attract soot particles*

*Electric wires give soot a negative charge*

*Black smoke enters bottom of chimney*

## SMOKE SCRUBBERS

Static electricity can be used to clean filthy smoke from chimneys. The smoke flows over electric wires, so the soot particles inside it pick up a negative electrical charge. Higher up the chimney, there are more wires with an opposite, positive electrical charge. The negative soot particles cling to these positive wires, cleaning the dirt from the smoke.

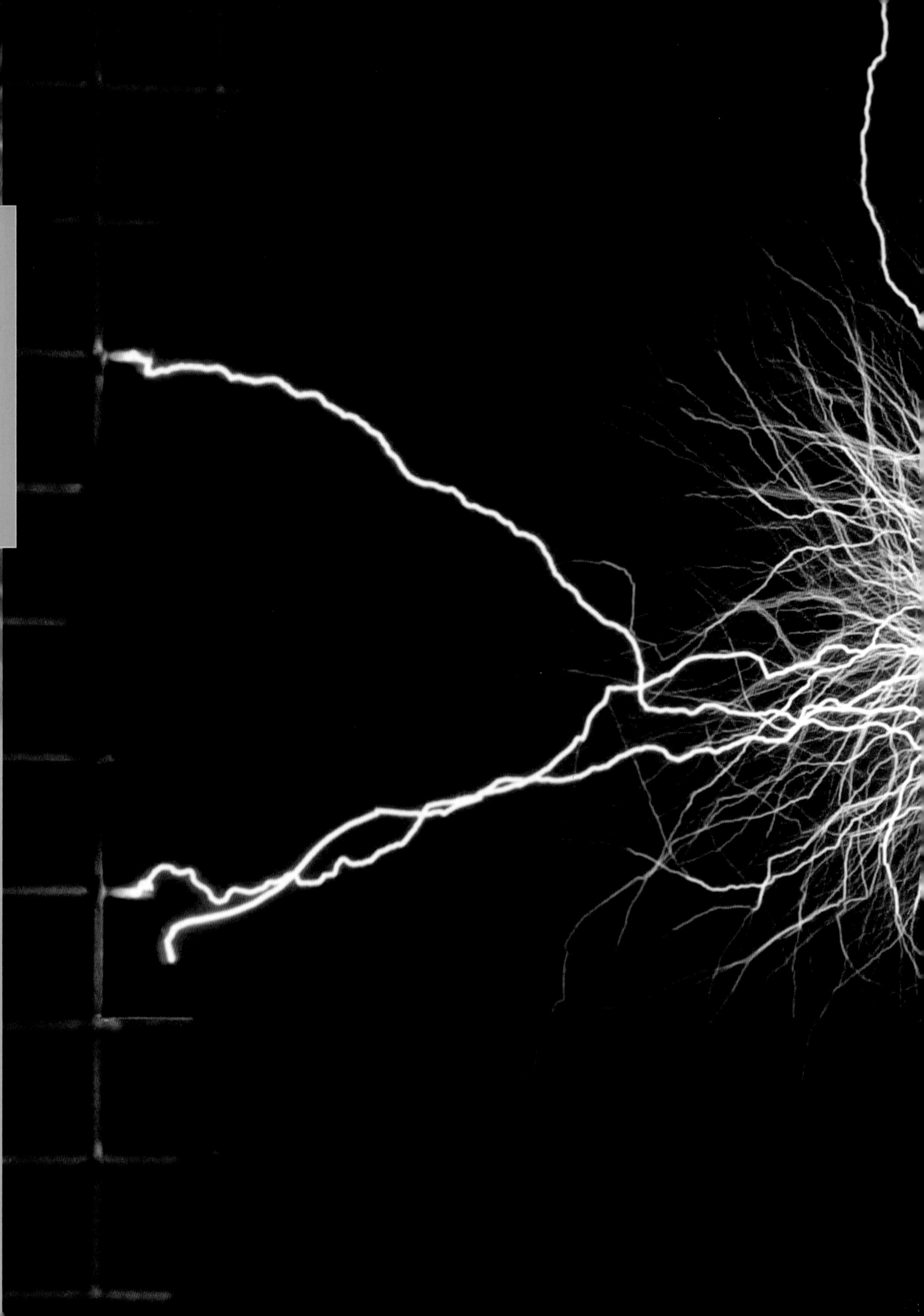

## LIGHTNING IN THE LAB

Scientists can create lightning in laboratories by allowing a huge electric charge to build up between a wire grid and an electrode (centre of the picture). Here, the lightning is being passed over a sheet of glass to test how a building would hold up to a lightning strike.

# Magnetism

Magnetism is a hidden force, caused by electrons inside the atoms from which all things are made. You stick to the ground partly because you are a little bit magnetic, and Earth is like a giant magnet. You do not notice the magnetism between Earth and your body because gravity is stronger. For small things, magnetism can be much stronger than gravity – that is why magnets cling to your refrigerator.

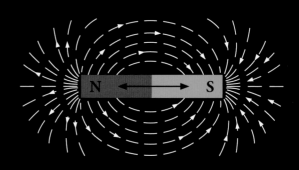

## MAGNETIC FIELD PATTERNS

Magnets have two ends – a north pole (N) and a south pole (S). A magnetic field (invisible pattern of magnetism) curves from north to south. Put a nail in this field and it is pulled towards the magnet.

## HOW DOES A MAGNET WORK?

Things that are magnetic, such as an iron bar, are made up of many tiny areas called domains. Each one is like a mini magnet. When the domains face in the same direction, the object is magnetized. When the domains are mixed up, it is demagnetized.

*Stroking a magnet across the bar makes the domains line up.*

*In an unmagnetized iron bar, domains are mixed up and face different directions.*

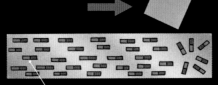

*Hitting a bar with a hammer mixes up the domains, so the magnetism disappears.*

*In a magnetized bar, domains line up and all face the same way.*

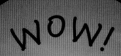

## WOW!

Some bacteria are magnetic. Scientists have used magnets to steer these bacteria around to build microscopic walls.

▼ NORTHERN LIGHTS *These strange green patterns swirl in the sky close to the North Pole. They happen when particles shooting from the Sun are pulled by Earth's magnetism, crashing into our atmosphere to make light.*

## MAGNETIC SWITCHES

Magnets can catch burglars. A magnet sits on the door, and a magnetic switch sits on the frame. When the door closes, the magnet pushes apart two contacts in the switch. When the door opens, the switch moves away from the magnet and the contacts spring together. This completes a circuit and the alarm rings.

*Magnetic contacts in the switch are pushed apart when the door is closed.*

*When the magnet moves away, the contacts spring back together.*

## HOW STRONG IS THAT MAGNET?

Magnets come in all sizes. We can make magnets out of ceramics or metal, such as the ones you find on your refrigerator. We can also create powerful magnets using electricity (see pp.184–185). Our bodies contain tiny amounts of iron, so we work like very weak magnets. The Earth is a giant magnet, but its magnetism is spread out through the whole planet. The strongest magnetic forces occur inside atoms, but only over very tiny distances.

**Magnet strengths compared to an ordinary bar magnet.**

| OBJECT | STRENGTH |
|---|---|
| Electromagnet | 500 |
| Ceramic magnet | 2 |
| Bar magnet | 1 |
| Human body | 1/50 million |

## EARTH THE MAGNET

Molten matter swirling inside Earth's core makes a magnetic field all around us. It is as though there is a huge but quite weak magnet hiding inside our planet.

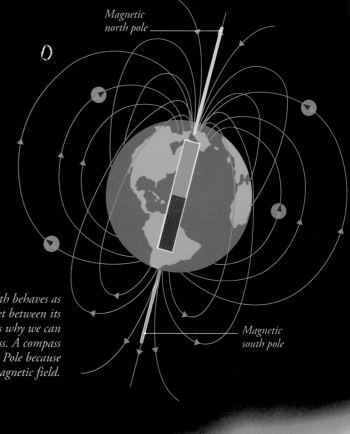

*Magnetic north pole*

*Magnetic south pole*

▶ FINDING YOUR WAY *Earth behaves as if there is a giant invisible magnet between its North and South Poles. That is why we can find our way using a compass. A compass points towards Earth's North Pole because it is attracted by Earth's magnetic field.*

# Electromagnetism

When your doorbell rings, the noise you hear is made by
a quick burst of magnetism and electricity. Many people
think electricity and magnetism are separate things, but
really they are like two sides of the same coin. We call this
electromagnetism, and it has important uses in our daily lives.

## WHAT IS ELECTROMAGNETISM?

Both electricity and magnetism are made
by the electrons inside atoms. When
electricity changes, it makes magnetism.
When magnetism changes, it makes
electricity. To make magnetism powerful
enough to lift things, you need lots of coils
of electric wire and a big electric current.

## HOW DO YOU MAKE AN ELECTROMAGNET?

When electric current flows down a
wire, it makes invisible magnetism all
around it. If you bend the wire into a
spiral and wrap it round a nail, the nail
becomes an electrically powered magnet –
an electromagnet. Real electromagnets
work just like this but on a bigger scale.

▶ SCRAPYARD SWING *This giant crane
lifts metal with an electromagnet. When the
electricity is turned on, the metal sticks.
When it is turned off, the metal falls away.*

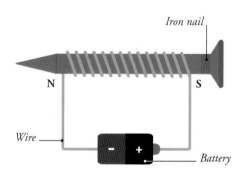

*Iron nail*

N        S

*Wire*

–   +

*Battery*

▲ SIMPLE ELECTROMAGNET
*You can make an electromagnet by
wrapping plastic-coated wire around a
nail and attaching it to a small battery.*

## WOW!

A tiny fridge magnet
is 20 times more
powerful than the
magnetism made by
our huge planet Earth.

*Electromagnet attracts
magnetic metal parts
when current is switched on*

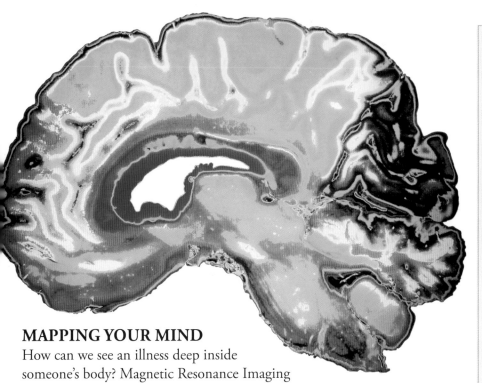

### JAMES CLERK MAXWELL

Electromagnetism was always there, waiting to be discovered. The first person to understand it properly was Scotsman James Clerk Maxwell. In 1865, he wrote four simple rules explaining how electricity could make magnetism and vice versa.

## MAPPING YOUR MIND

How can we see an illness deep inside someone's body? Magnetic Resonance Imaging (MRI) scanners use electromagnets to excite the body's atoms. This makes them form magnetic patterns that a computer can see and draw on a screen. Doctors can examine these scans to pinpoint diseases and save lives.

▲ BRAIN SCAN *This is what the cross section of a person's brain looks like through an MRI scanner.*

## YOUR ELECTROMAGNETIC HOME

There are dozens of electromagnets in your home. They are in your doorbell and your telephone. They are in your radio and television, and in the loudspeakers and microphones in your stereo. You will also find them in low-energy light bulbs, battery chargers, and burglar alarms. In fact, they are all around you.

▶ ELECTRIC DOORBELL *Two orange coils of copper wire turn into electromagnets when electricity flows through them. They pull on the clapper that rings your bell.*

*Bell rings when clapper strikes it*

*Copper coils become electromagnets when electricity flows*

*Magnetized coils pull bell clapper towards them, then release it*

*When you push a switch, wires carry electricity to copper coils*

# Electric motors and generators

When electrons pass through a wire, they can power anything from a toy car to a huge train. Electric motors hold the secret to this trick – they turn electrical energy into kinetic energy (pp.128–129). Machines called generators work the opposite way to motors. They turn kinetic energy into electricity. All of the electricity we use comes from generators.

## WHAT IS AN ELECTRIC MOTOR?

An electric motor is a loop of wire sitting inside a piece of magnetic metal (a permanent magnet). When the wire is connected to an electricity supply, a magnetic field forms around it. Force from the permanent magnet pushes against the magnetized wire, spinning it round.

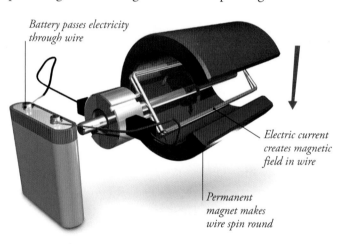

*Battery passes electricity through wire*

*Electric current creates magnetic field in wire*

*Permanent magnet makes wire spin round*

## WHAT IS A GENERATOR?

Electricity generators are almost identical to motors but work in exactly the opposite way. If you spin a piece of wire inside a magnet, electricity flows through it. The longer you spin the wire, the more electricity you can make.

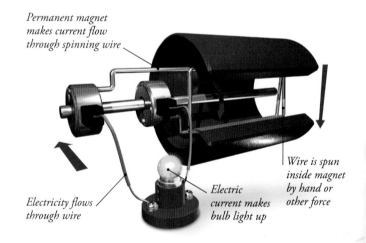

*Permanent magnet makes current flow through spinning wire*

*Electricity flows through wire*

*Electric current makes bulb light up*

*Wire is spun inside magnet by hand or other force*

## LINEAR MOTOR

Electric motors make things turn in circles, so they are perfect for powering washing machines or food mixers. Some machines need to go in straight lines, not circles, so they use a linear motor instead. This is like a normal motor cut open and laid out flat. One part of the motor slides past the other, gliding down a long, straight track.

▶ ROCKET LAUNCHER *In the future, linear motors could fire rockets into space. This test shows a small model rocket on a linear motor track.*

## PEDAL POWER

Dynamos are like mini power stations strapped to your bike. As you cycle along, your tyre presses against a tiny wheel on the dynamo, spinning it round. Inside the dynamo, there is a small electricity generator. This makes enough electric current to feed the bulb in your headlight.

*5. Bulb lights up*

*4. Cable carries electricity to bulb*

*1. Wheel spins round*

*2. Dynamo wheel is driven by tyre*

*3. Generator inside dynamo makes electricity*

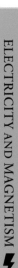

## PUMPING OUT POWER

Hydroelectric power stations use the force of a rushing river to make electricity. The river is blocked by a vast concrete dam, then slowly allowed to escape through a pipe at the bottom. The pipe fires the river water at a windmill-like propeller, called a water turbine. The turbine spins a huge generator and produces vast amounts of electricity.

▼ WATER POWER *The Hoover Dam in the United States uses 11 giant water turbines and generators to make as much electricity as a large coal or nuclear power station.*

## WOW!

The Hoover Dam on the Colorado River, in the United States, can make as much power as half a million bicycle dynamos.

# Using electricity

In ancient times, people had to burn substances such as wood or coal to get heat and light energy for cooking and keeping warm. In our modern homes, we can just flick a switch to release electrical energy whenever we need it. Our homes are buzzing with electricity. Running through the walls and the ceiling, hidden wires conduct power to every room.

## HOW DOES AN ELECTRIC LIGHT WORK?

Old-fashioned lightbulbs create light by making heat. They squeeze a large electric current through a narrow, twisted wire called a filament. As the current flows, the filament gets so hot that it glows brightly and gives off light. This wastes a lot of energy as heat. More modern lights save energy by using cold, fluorescent tubes and light-emitting diodes (LEDs).

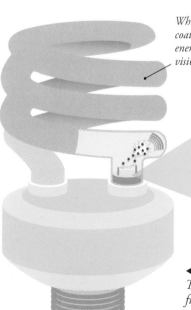

◄ FILAMENT LIGHT *The filament is made of hard tungsten metal. It is covered by a glass bulb filled with argon gas, to stop it burning so it lasts longer.*

*White phosphor coating turns energy into visible light*

*Tube is filled with mercury vapour*

*Electrons release energy when they hit mercury vapour*

*Individual LED in plastic case*

*Electrode fires electrons through the tube*

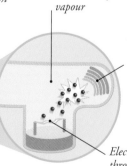

◄ COMPACT FLUORESCENT LAMP (CFL)
*These clever bulbs use four times less energy than filament lamps and last 10 times longer. They work by firing electrons through a tube filled with mercury vapour to make light.*

▲ LIGHT-EMITTING DIODE (LED)
*These use half as much energy as CFL bulbs and last twice as long. LEDs pass electrons through crystals to make light.*

## MAKING HEAT WITH ELECTRICITY

When electricity flows through wires, it produces heat. "Heating elements" are wires designed to get as hot as possible when electricity passes through them, and are used in devices such as kettles and hair dryers. They are made from special alloys to stop them burning. As the current flows in the wire, it shakes the atoms inside. The shaking atoms crash about and radiate heat.

▲ INSIDE A TOASTER *The heating elements in a toaster get hot as electricity passes through them. They radiate heat towards the bread, turning it into toast.*

## WIRED FOR POWER

You live inside giant electric circuits. Every room in your home has wires running above, below, and right around it. Homes often have many separate circuits. A high-powered circuit carries a large electric current to the cooker. A separate lighting circuit powers all the lights and a power circuit carries current to all other appliances.

◄ HOUSEHOLD WIRING *Separate circuits power different appliances in your home. Each circuit is protected by a separate fuse (safety switch). If one circuit fails, the others keep working.*

**KEY**
▬ *Lighting circuit*
▬ *Power circuit*

## WHY DO COMPUTERS GET HOT?

Computers work by using electronic circuits to shuffle numbers around. Although they have few moving parts, they still get hot. Every time a computer chip changes a number, it has to use electrical energy to do it. This releases a tiny bit of heat. Even the simplest job a computer does changes millions of numbers, so the heat quickly mounts up.

► HOT LAPTOP *This coloured photograph shows the temperature of a laptop. The hottest bit is the processor (red). The coolest bit (blue) is the screen.*

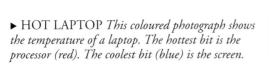

# Electricity supply

Every time you switch on a lamp or plug in your vacuum cleaner, electrons charge into your home, hauling energy down the wires. That energy might have come from a power station far away, or it could have been absorbed by a solar panel on your roof. There are many different ways of making, storing, and using electricity.

## POWER PIONEER

Before electricity supply was invented, if people wanted heat in their homes, they had to burn wood or coal. If they wanted light, they needed candles or oil lamps. In 1882, US inventor Thomas Edison (1847–1931) opened the world's first proper power station in Manhattan, New York. Though it only supplied 59 homes to begin with, it proved the value of electricity and changed the world.

▲ PEARL STREET *Edison's power station filled two large buildings on Pearl Street, Manhattan. Coal was burned in furnaces on the ground floor, powering six electricity generators on the first floor.*

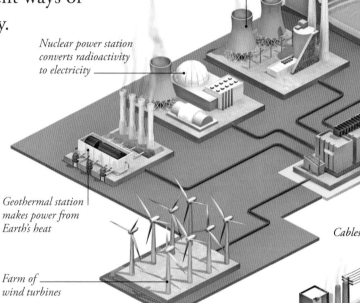

*Hydroelectric power makes electricity from river dam*

*Coal power station*

*Nuclear power station converts radioactivity to electricity*

*Geothermal station makes power from Earth's heat*

*Farm of wind turbines*

*Cables*

*Large factories use as much power as a small town*

## ELECTRICITY ON THE MOVE

A complex grid network of power lines moves electricity from places where it is made to places where it is used. The electricity that arrives at your home is exactly the same, whether it has come from coal, wind, or sunlight. There are many different ways of making electricity, and we can use it for many different purposes.

## VOLTAGE CHANGE

Voltage is the force that pushes electricity down wires. High voltage is better for transferring electricity over long distances, because it uses less energy. In the home, we use lower voltages because they are safer, but things that use high power, such as air conditioners and refrigerators, usually need high voltages as well. Batteries give out low power with smaller and safer voltages.

**Lightning bolt 100 million volts**

**Pylon 500,000 volts**

**Electric train cables 20,000 volts**

## STORING ELECTRICITY

Many devices, such as mobile phones, have batteries that can store electricity, so we can use them when we are away from home. Rechargeable batteries take in electricity from the mains supply whenever they are plugged in, storing it as potential energy. They release this energy as electricity when the phone is used.

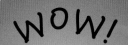

**WOW!**

Worldwide, we use more than 40 billion disposable batteries every year – enough to stretch to the Moon five times if you lined them up.

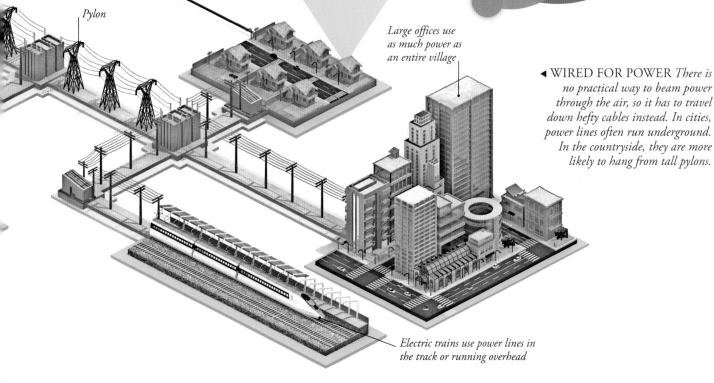

*Solar panels convert sunlight into electricity*

*Rooftop solar panel supplies home with most of its own power*

*Pylon*

*Large offices use as much power as an entire village*

◄ WIRED FOR POWER *There is no practical way to beam power through the air, so it has to travel down hefty cables instead. In cities, power lines often run underground. In the countryside, they are more likely to hang from tall pylons.*

*Electric trains use power lines in the track or running overhead*

**Electric eel**
**600 volts**

**Electricity in homes**
**110–240 volts**

**Laptop charger**
**20 volts**

**Torch bulb**
**1.5 volts**

# Energy sources

A billion cars, six billion mobile phones, a billion televisions – it is no wonder Earth's seven billion people use vast amounts of energy. Most of this energy is beamed down to us from the Sun. Some has been stored inside the Earth for millions of years as fossil fuels, which we can burn to make power. Because this causes environmental problems, people are now looking to cleaner, "greener" forms of energy, such as wind and solar power.

## NATURE'S POWER SOURCE

Energy from the Sun powers most of the activity on our planet, from plants and animals to wind and weather. Deep inside the Sun, tiny atoms smash together and fire off energy. About 150 million km (93 million miles) away, here on Earth, sunlight makes plants burst to life, feeding animals, including people. A single second of the Sun's fiery energy could power everything that happens on Earth for a million years.

## POWER FOR HUMANS

Humans use energy to power machines such as cars, computers, and factories. About 80–90 per cent of our energy comes from fossil fuels: petroleum (oil), coal, and natural gas. Most renewable energy comes from hydroelectricity (river dams) and biomass (plants burned to make power). Only a tiny bit comes from solar power and wind power.

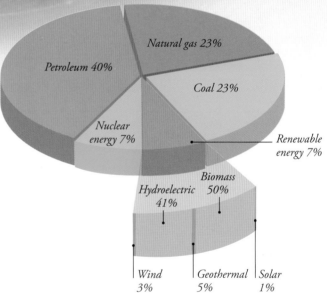

Natural gas 23%

Petroleum 40%

Coal 23%

Nuclear energy 7%

Renewable energy 7%

Hydroelectric 41%

Biomass 50%

Wind 3%

Geothermal 5%

Solar 1%

▲ COAL MINING *Coal is extracted from deep underground with the help of big machines that drill and blast the surface of Earth.*

## BURIED FUELS

Oil, coal, and gas are made from fossils. When plants and sea creatures die, their remains get buried, squashed down, and cooked by Earth's heat. After many millions of years, this creates fossil fuels we can dig out or pipe to the surface.

## NUCLEAR ENERGY

When small atoms smash together or big ones break apart, they release energy trapped inside. This comes from the nucleus (centre) of the atoms, so it is called nuclear energy. Nuclear power stations can make lots of electricity, but they also create dangerous waste.

▶ NUCLEAR POWER *Nuclear plants need vast amounts of cooling water, so they are often built close to the sea.*

◀ SOLAR PANELS *These sun collectors make electricity when light falls on them. The light pushes electrons through a circuit, making the electricity flow.*

## ENDLESS ENERGY

Energy that we can make forever, without burning fossil fuels, is called renewable energy. Solar power, wind power, biomass, and energy from rivers and oceans are all renewable. Very little of the energy we use is renewable at the moment but, in the future, more of our energy will need to be made this way as fossil fuels become scarce.

**WOW!**

It takes about 1,500 wind turbines to make as much power as a big coal or nuclear power station.

### CLIMATE CHANGE

Burning fossil fuels release a gas called carbon dioxide. This is building up around Earth, trapping heat and warming our planet. As Earth gets hotter, its climate is changing, and the seas are slowly rising. These things will make it harder for millions of people to feed themselves and live safely.

Heat from the Sun warms Earth

Some heat from Earth escapes

Heat-trapping gases surround Earth

Some heat from Earth is trapped by atmosphere, warming planet

▼ ECO HOMES *The BedZED building in London, UK, has wind cowls on the roof so it can keep cool without air conditioning.*

## MAKE YOUR OWN POWER

Ordinary power stations waste up to two-thirds of the fuel they burn. Homes that make their own power can be more efficient. Many homes now have solar panels or mini wind turbines on their rooves. Others use heating boilers that burn biomass (wood pellets) instead of gas and oil.

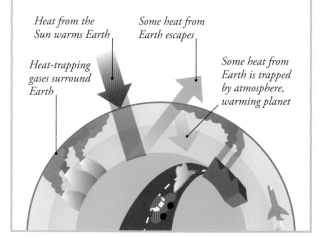

## SOLAR POWER PLANT

The Sun's energy could replace polluting fossil fuels as an electricity source, if we could only find efficient ways to capture it. This solar power plant in California, USA, used hundreds of mirrors to focus sunlight on a central tower, which made intense heat and generated electricity.

# Electronics

Electronics is a kind of "thinking" electricity. While we use electricity to power things such as lights and motors, we use electronics to control electrical things. Where electricity uses lots of current to deliver a large amount of energy, electronics uses only tiny amounts. Electronic circuits use lots of tiny switches to change the flow of electrons. Some electronic parts are so sensitive they can move just a single electron.

ELECTRICITY AND MAGNETISM

## ALL-IN-ONE CHIP

Electronic gadgets, such as mobile phones, have complex circuits inside them. Each circuit is made from thousands, millions, or even billions of separate parts. As they would take up a lot of room, engineers had to find ways of squeezing these parts into a smaller space. This led to the invention of the integrated circuit, or microchip – a flat, fingernail-sized circuit with electrical parts made in miniature form.

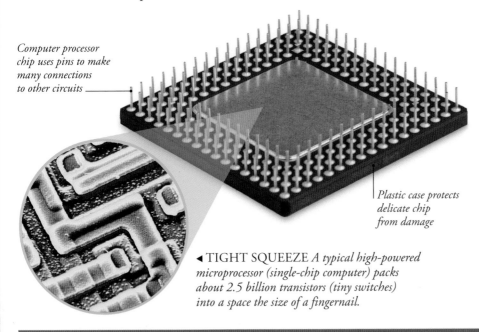

*Computer processor chip uses pins to make many connections to other circuits*

*Plastic case protects delicate chip from damage*

◀ TIGHT SQUEEZE *A typical high-powered microprocessor (single-chip computer) packs about 2.5 billion transistors (tiny switches) into a space the size of a fingernail.*

### TRANSISTOR

American scientists John Bardeen, William Shockley, and Walter Brattain invented the transistor in 1947. A transistor is the king of electronic parts – the most important one by far. It is a simple, miniature on/off switch that can also make an electric current much bigger. If you connect lots of transistors together, you can build computer circuits that make decisions and remember things.

## ELECTRONIC BUILDING BLOCKS

Electronic circuits are made by connecting small parts called components. Each one does a different job and they all work as a team. A typical electronic circuit in a transistor radio would have a few dozen components. The processor chip in a computer could have billions.

▲ DIODES
*Diodes make electricity flow in only one direction. They are used for picking up signals in radios.*

▲ LEDs
*Glowing lights on electronic gadgets are made from tiny coloured light-emitting diodes (LEDs).*

▲ TRANSISTORS
*Transistors switch currents or make them bigger. They were first used to make sounds louder in hearing aids.*

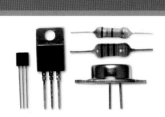

## HIDDEN CHIPS

Shops hide microchips inside items to stop people stealing them. The chips pick up radio signals from plastic gates standing on either side of the shop doors. When an item is bought, the sales assistant switches off the chip so it passes through the gates harmlessly. If something is stolen, the chip sends a signal to the gates and sounds the alarm.

*A miniature circuit picks up a radio signal from gates by the shop's doorway.*

▲ SECRET CIRCUIT *This anti-shoplifting chip is called a radio-frequency identification (RFID) tag. It has an electronic circuit hidden inside.*

*Flexible Organic LED (OLED) displays can bend into new shapes without breaking.*

## FLEXIBLE GADGETS

The first radios and TVs were huge, heavy, and gobbled up power. The first computers filled entire rooms. Today's electronic gadgets are small and light – we can easily carry phones in our pockets. Tomorrow's gadgets will be even more user-friendly. They will be made from very light and flexible plastic circuits with bendy displays, known as Organic LEDs (OLEDs). In the future, your phone could wrap around your wrist like a watch.

▲ RESISTORS
*A resistor makes an electric current smaller. Dimmer switches use resistors to turn down lights.*

▲ CAPACITORS
*A capacitor stores electricity, like a battery without chemicals. The flash on a camera uses a large capacitor.*

## WOW!

You could fit more than 2 million tiny transistors on the full stop at the end of this sentence.

# Radio and television

A television can conjure up people from history or show you what is happening right now on the other side of the world. This might seem like magic, but it is science in action. A TV set grabs invisible signals shooting down cables, or through the air, and converts them into pictures you can see. Radio does exactly the same thing but with voices and music.

## WHAT IS A RADIO WAVE?

Radio and TV signals beam through the air as waves of energy. These radio waves are invisible, yet they can pass through buildings. They travel as fast as light waves, so they can speed seven times around the world in a single second. If you could see a radio wave, it would have a snake-like pattern of ups and downs.

Time

High-frequency radio wave

Volume

Wavelength

Time

Low-frequency radio wave

Volume

Wavelength

▲ RADIO WAVE *Radio and TV signals travel in waves of different frequency. The frequency is how many times the wave goes up and down each second. You can pick up different channels by tuning to different frequencies.*

## RADIO BY NUMBERS

Modern radio stations use digital signals to transmit sounds. The transmitter breaks the signal from a radio station into tiny chunks. Each chunk is turned into a number and beamed through the air many times. A digital radio picks up the chunks and turns them back into waves you can hear. This makes a clear sound with no crackle.

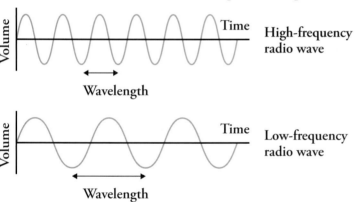

Radio station turns signal into numbers

Building

**WOW!**

By bouncing signals from satellites, we can send TV and radio transmissions to anywhere in the world.

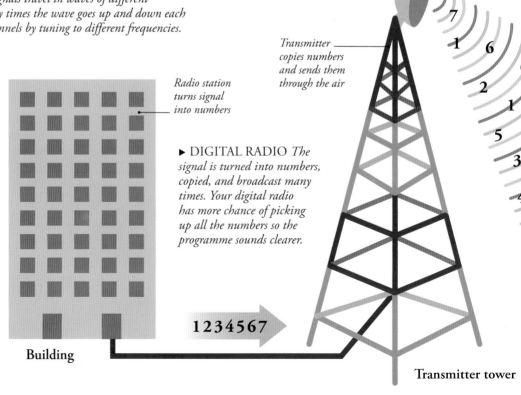

Transmitter copies numbers and sends them through the air

▶ DIGITAL RADIO *The signal is turned into numbers, copied, and broadcast many times. Your digital radio has more chance of picking up all the numbers so the programme sounds clearer.*

1234567

Transmitter tower

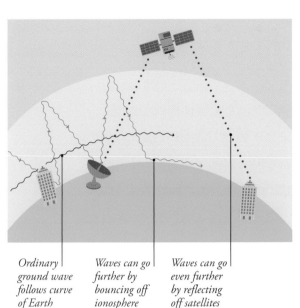

Ordinary
ground wave
follows curve
of Earth

Waves can go
further by
bouncing off
ionosphere

Waves can go
even further
by reflecting
off satellites

## STREAMING SIGNALS

Radio waves travel in straight lines, so you might think they would shoot into space and disappear. In fact, radio signals can bounce right around our curved planet. Some crawl across the ground. Others go further by bouncing off part of the atmosphere called the ionosphere. This works best at night, which is why you can hear distant radio stations better in the evening. Space satellites help signals go even further by bouncing them like mirrors in the sky.

*Early TVs could only show pictures in black and white.*

## HOW TV MAKES HISTORY

Television helps us mark important moments in history. Thanks to satellites, we can watch amazing events as they happen ("live") anywhere in the world. By storing TV pictures as videos, we can keep these moments forever. Recording history like this was impossible before TV was invented.

▲ MOON LANDING *When astronauts reached the Moon in 1969, up to a billion people followed the event on radio and TV.*

*Each part of a TV picture is built from tiny red, blue, and green dots called pixels.*

◄ COLOUR SCREEN *Digital televisions receive signals with streams of numbers and turn them into pictures and sound.*

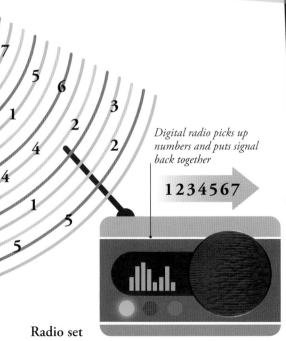

*Digital radio picks up numbers and puts signal back together*

1234567

**Radio set**

## HOW TELEVISION WORKS

Television works just like radio except the waves travelling through the air carry pictures as well as sounds. When the signal streams into your home, your television splits it in two. Part of it makes the picture and the other part makes the sound. By sending their signals as numbers, just like the digital radio, TV stations can send lots of channels at once, giving us more choice of things to watch.

# Computers

Most animals think with their brains, but people also think with electronic brains – computers. A hundred years ago, there were no electronic computers anywhere in the world. Today, there are more computers than people. They hum away on office desks and control robots in factories. In our homes, they power everything from phones and TVs to washing machines and microwaves.

## HISTORY OF COMPUTERS

No single person invented the computer. The idea took shape gradually, starting with the abacus, a simple counting frame made of beads, invented about 4,500 years ago. Although modern computers do much more complicated tasks, they are still really just counting machines.

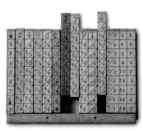

▲ NAPIER'S BONES *The first computers used moving rods, levers, or gears. This simple wooden multiplier was invented by Scotsman John Napier in 1617.*

▲ ENIAC *The world's first electronic computer was invented in 1946. It was as big and heavy as a truck, but less powerful than a modern mobile phone.*

▲ APPLE MAC *This machine appeared in 1984. With a friendly picture-based screen, it was designed for people who had never used a computer before.*

▶ TABLET *Larger computers use keyboards and mice, but tablets do not need them. Instead of typing words or dragging your mouse, you simply touch the screen with your finger.*

*A sensitive surface hidden inside the touch screen uses electricity to detect your finger.*

## HOW DOES A COMPUTER WORK?

From editing holiday photos to playing music videos, everything computers do happens in four steps. First, we feed in (input) some information, often with a keyboard or mouse. Next, the computer stores that information in its memory. Then it works on the information with its processor chips, doing millions of calculations each second. Finally, we get the results (output) on the screen.

*Each key is a tiny switch that sends an electric pulse to the computer.*

**Input devices**

**Output**

*A computer screen is similar to a TV, but designed for close-up viewing.*

*Hard drives store information on magnetic plates, spinning at high speed.*

**Storage**

**Processor**

*A computer's "brain" is an electronic chip the size of a fingernail.*

## WOW!

Today's supercomputers are about 300 trillion (300,000,000,000,000) times faster than the 1940s ENIAC.

*The electronic parts in a tablet computer are on very flat circuits, not much thicker than cardboard.*

## EYES ON THE ROAD

People cannot always reach for a computer to look up the facts and figures they need. Heads-up displays project useful information before your eyes. Drivers can have maps or information from the dashboard beamed onto their windscreen so they do not have to look away from the road.

▼ MORE SPEED *NASA's Pleiades supercomputer has 112,896 processors stacked in 185 racks, and is used for space and weather research.*

## SUPERCOMPUTERS

Forecasting the weather is a tricky scientific problem that would take an ordinary computer years to tackle. Supercomputers work millions of times faster. They are made by connecting hundreds of thousands of ordinary computer chips together. A problem is broken up into tiny pieces and each chip handles one small piece. This is much quicker than trying to tackle the problem all together, in one big piece.

# Mobile devices

If you could see a mobile phone call, it would look like waves of energy spreading out from the phone to a tower or satellite. That is because mobile phones connect together with radio waves travelling at the speed of light. They do not need wires, so we can use them anywhere in the world, from the top of Mount Everest to the middle of the Sahara desert.

## HISTORY CALLING
Mobile phones were invented by a US engineer, Martin Cooper, in 1973, but it was another 10 years before they went on sale. The first mobile phone you could buy was the Motorola DynaTAC, which cost US$3,995. It was six times heavier than a modern phone, and its huge battery lasted only an hour.

## HOW MOBILE PHONES WORK
Mobile phones make calls by sending signals through the air using radio waves. Mobile phone networks divide cities into much smaller zones called cells. Each cell contains its own mobile phone mast (sending and receiving antenna). The signals from a phone travel via the closest mast, which beams them on through the telephone network to their destination.

> ### FAST FACTS
> ■ There are 3.4 billion mobile phone users worldwide.
> ■ Five times more people have mobile phones than ordinary (landline) phones.
> ■ Mobile broadband (using the Internet on mobile phones) is the fastest-growing technology in history.

*When A calls B, the signal travels to the closest mast. This sends the call to the mast in the cell where B is, and to B's phone.*

*The cells are small in cities but larger in the countryside.*

*When you drive around, masts in nearby cells send silent signals to your phone, so the network always knows where you are.*

▲ PATCHWORK OF CELLS *Mobile phones operate within a network of cells. Cells vary in size and overlap slightly, so that when you move from one cell to another your call can be passed from one cell to the next.*

## CALLING BY NUMBERS

When you speak, your voice wobbles through the air in sound waves. When you talk into a mobile phone, it converts the sound of your voice into numbers (digits). This kind of digital information is easier and safer to send over networks, and travels more clearly with less background noise.

*Voice travels to mobile phones as sound waves*

10110011101010101
10100101010101010
10101010101010101
1111010101010101

*Sound waves converted into digits*

*Mobile phones send and receive calls as streams of digits*

## WOW!

The first mobile phone invented was as big as a brick and weighed 13 kg (28 lb) – 100 times more than a modern phone.

## MOBILE WORLD

Mobile phones have really changed the world. It is much too expensive to lay giant telephone cables across huge continents such as Africa and Asia. Fortunately, mobile phones can work anywhere, without wires. Thanks to mobile phones, people can enjoy the Internet and stay in touch with their friends even in areas without telephone cables.

◄ CALLING FROM ANYWHERE
*Mobile phones can make calls wherever they can pick up a signal from a mast or overhead satellite – even a remote desert. The closer they are to the mast, the more reliable the call.*

# The Internet

You can send messages to your friends at the click of a button, even if they live on the other side of the world. This is possible thanks to the Internet, a giant network that links the world's computers. The Internet is also the power behind the World Wide Web, a digital library made from more than a trillion pages that anyone can tap into anytime, anywhere.

## HOW DOES THE INTERNET WORK?

Just as there are many ways to travel between two places on Earth, so there are many ways to send information between two computers. The Internet uses this idea to transport data (computer information) very efficiently. The data is split into packets (tiny pieces), which travel separately so they reach their destination faster. Any information can be sent over the Internet, from photos and videos to emails and phone calls.

▶ PACKET SWITCHING *When you email a photo to a friend, it is broken into thousands or millions of packets. Each packet might travel by a very different route to the same destination.*

*1. A photo is broken into thousands of small packets.*

*2. Each packet is turned into a series of numbers and labelled with the destination address.*

*3. Packets travel by the quickest route, but not all together.*

## COMPUTER NETWORK

The Internet is a computer network that spans the world, but there are smaller networks too. A network at home may connect several computers to a printer. This is called a LAN (Local Area Network). In offices, there are bigger LANs, where dozens of computers are connected. Each LAN usually has one place where it connects to the outside Internet, through a secure connection.

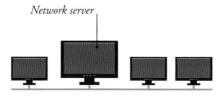

*Network server*

▲ LINE NETWORK *Each computer is connected to the others in a simple chain. The network server controls the network.*

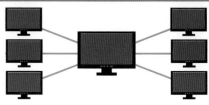

▲ STAR NETWORK *All the computers on this network branch out from the server in the middle, which controls them.*

## UNDERWATER INTERNET

The Internet links about 200 countries with enormous cables, almost 250 of which stretch under the sea. Most run from the United States, some from the east coast to the United Kingdom and Europe, and others from the west coast to China, Japan, and other Asian countries.

▶ CABLE CRAWLER *Most Internet cables are lowered from reels behind giant ships. Some are laid by submarines like this, which crawl along the seabed.*

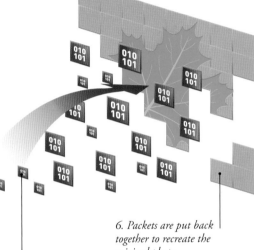

*6. Packets are put back together to recreate the original photo.*

*5. Packets eventually all arrive at the same place.*

*4. If any route is blocked, packets can go a different way.*

*Servers are made from individual computers slotted into racks in large cabinets.*

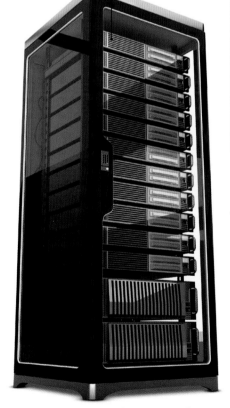

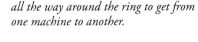

▲ RING NETWORK *Data has to travel all the way around the ring to get from one machine to another.*

▲ WEB SERVER *Websites operate from large computers called web servers, many times more powerful than home computers.*

### TIM BERNERS-LEE

English computer programmer Tim Berners-Lee invented the World Wide Web (WWW) when he saw how difficult it was for scientists to share their ideas. Instead of profiting from his idea, he has allowed people to use it for free.

## WORLD WIDE WEB

Knowledge used to be locked in libraries. Now you can read it from your computer on a giant library called the World Wide Web. Most web pages are stored in a simple format called HTML, which all computers recognize. Computers send and receive HTML pages using a method called HTTP. Simply speaking, the Web works because all the computers connected to it speak the same language of HTML and HTTP.

ELECTRICITY AND MAGNETISM

205

# Robots

As technology advances, more and more robotic machines are being invented to do jobs for us. Powered by electricity, they never get tired. Robots probably built your car, cut and stitched your clothes from the fabric, and put together many of the things you use each day.

## WHAT IS A ROBOT?

A robot is a machine that can learn to do different jobs. Controlled by a computer "brain", it is built from levers and wheels, and operated by electric motors. Some robots look like people, but most are just mechanical arms with tools attached.

▲ FACTORY FLOOR *Factory robots work 24 hours a day doing the same job without making a mistake. Robots can easily be reprogrammed to do new things.*

*Grabber arm can lift objects*

*Camera sends images back to the operator*

## WOW!

There are 1.5 million factory robots currently at work worldwide. Forty per cent of them are used to build cars.

## EMERGENCY ROBOTS

Robots can be sent on emergency missions that are too dangerous for people. Robots are now being developed that will be able to rescue casualties from earthquakes and other disasters. One day these robots will be able to dig through rubble, put out fires, and even drive cars.

◄ REMOTE CONTROL *Most robots do not control themselves. This bomb disposal robot, called Hobo, is operated from a distance by a person who uses the small camera on top to see what the robot is doing.*

## THINKING MACHINES

Humans learn how to do new things by thinking about the way they have done things before. This is called intelligence and it includes learning from your past mistakes. In the future, computers and robots will be artificially intelligent – as clever at thinking as humans.

▲ GRAND MASTER
*Computers are now smart enough to beat humans at games. In May 1997, IBM's Deep Blue supercomputer beat Russian world chess champion Gary Kasparov. In 2011, another IBM computer won the American TV quiz show* Jeopardy!*.*

## ROBOTS LIKE HUMANS

Robots need to work with humans, and scientists are always looking for ways to make them more like us. The latest robots even look like friendly humans. In the future, humans and robots might even merge. People can already have replacement robotic body parts and, in time, our entire bodies might be replaced with robotic ones.

## OUR ROBOT FUTURE?

If robots get too clever, they might want to control themselves instead of listening to us. American author Isaac Asimov suggested three laws that robots must obey: robots must not hurt people, they must always obey human orders, and they must protect themselves unless that involves harming people.

▲ I, ROBOT *This 2004 science fiction film is about a future world where most things are done by robots. It is based on a 1950 book of short stories written by Isaac Asimov.*

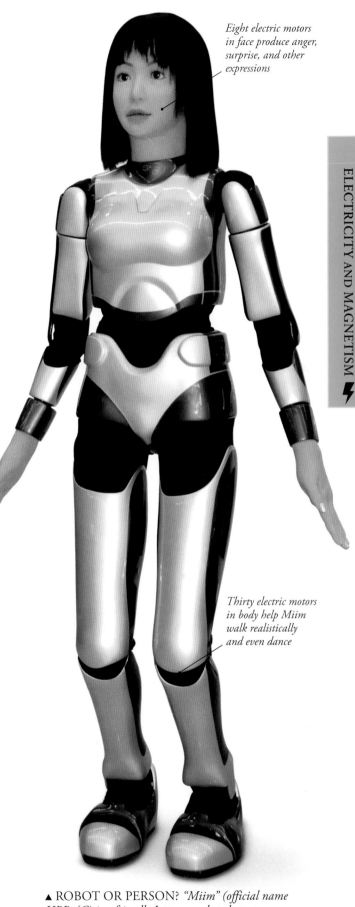

*Eight electric motors in face produce anger, surprise, and other expressions*

*Thirty electric motors in body help Miim walk realistically and even dance*

▲ ROBOT OR PERSON? *"Miim" (official name HRP-4C) is a friendly Japanese robot that can speak, sing, and model clothes on a catwalk. Robots made to look like humans are called androids.*

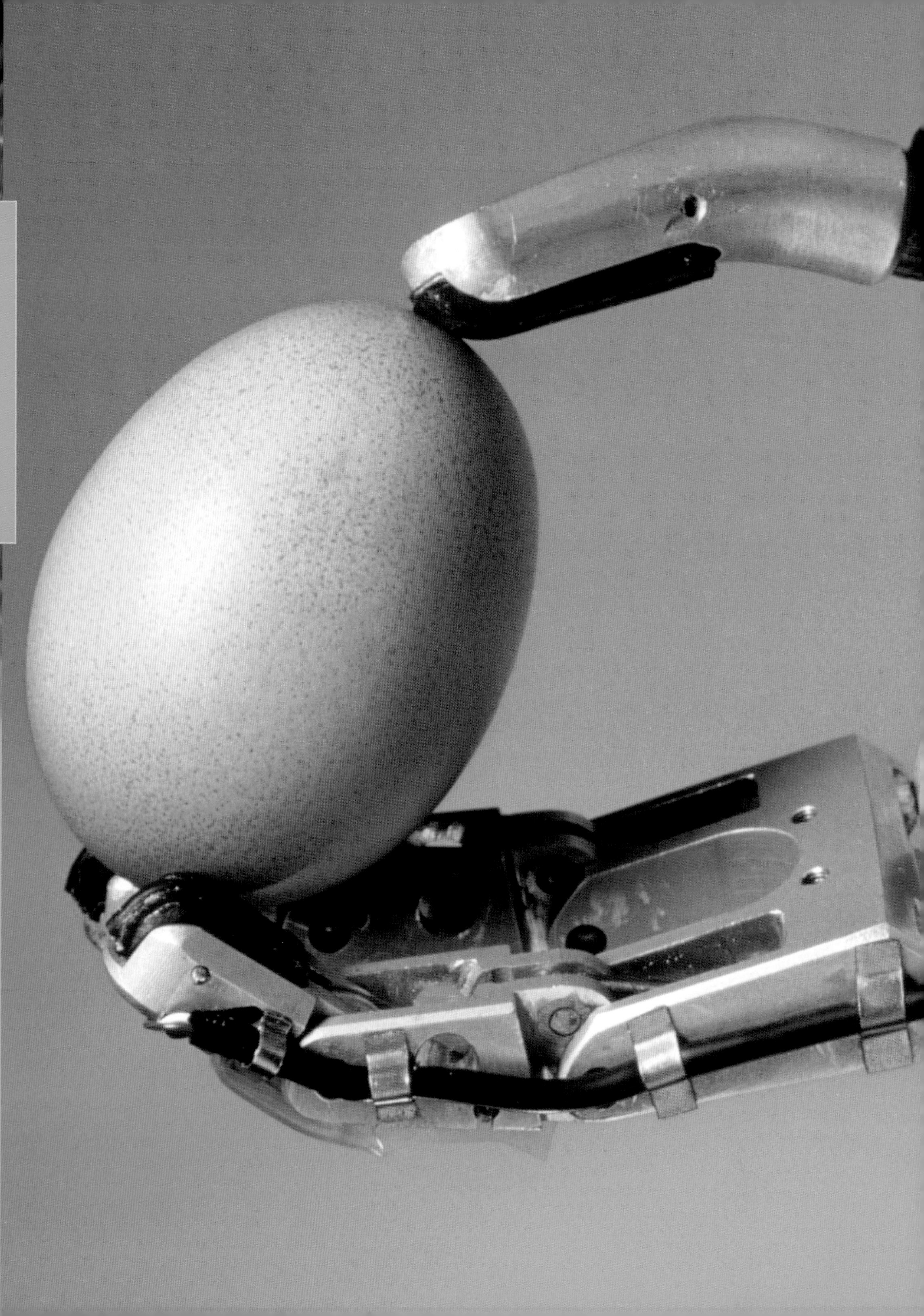

## ROBOT HAND

This robotic hand is strong enough to pick up heavy objects, but delicate enough to hold an egg without breaking it. It was designed for people who have lost a hand. The robot replacement can turn and grip, and has sensors on the fingers to make sure it does not squeeze too hard.

# Future technology

The new ideas and inventions of today could transform everyday life in the future. Electric vehicles gliding down the street, tiny robots zooming through our blood, endless energy captured from atoms – all this new technology could arrive in our lifetime. People have been making amazing inventions since the beginning of history. For thousands of years, technology has helped us to make the world a better place.

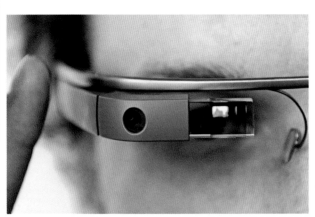

## HELPING HANDS

In the future, computers and machines will be with us all the time. We might wear glasses that tell us where we are so we never get lost, or clothes that transform depending on the temperature so we never get too hot or too cold.

◄ GOOGLE GLASS *These futuristic goggles project information onto a miniature computer screen right in front of your eyes.*

## BUILDING WITH ATOMS

The only materials we have today are ones we can find buried underground or the ones we design for ourselves in laboratories. One day soon, we will be able to make any material we like by building it out of atoms and molecules. This is called nanotechnology.

▼ NANOBOTS *It is hard to cure serious illnesses without risky surgery. In the future, we could swallow nanotechnology robots to repair our bodies from the inside.*

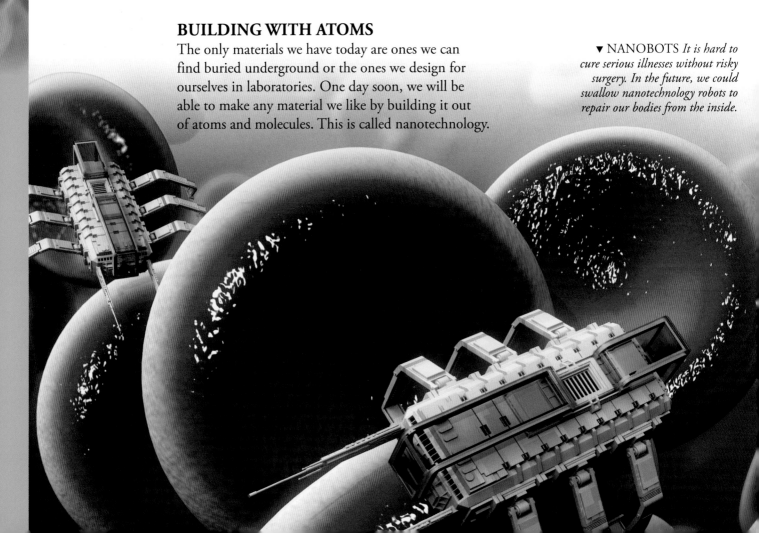

## CLEAN, GREEN MACHINES

There is only one planet Earth, so we must look after it. We need to use energy more wisely and make less pollution. One way to do this is to stop powering cars, ships, and planes with fossil fuels and use more electricity-powered vehicles. Electric vehicles are much "greener" because we can make the electricity using clean solar, wind, or water power.

▶ SOLAR BOAT *Instead of a diesel engine, this boat is powered by giant solar panels fixed to its roof. They drive an electric motor that spins the propellers.*

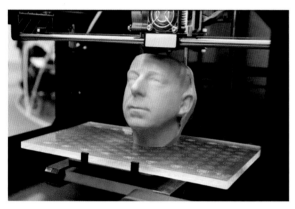

▲ GETTING AHEAD *This 3D printer is making a model of someone's head by slowly printing layers of melted plastic on top of one another.*

## MAKE IT YOURSELF

You can buy all kinds of things in shops, but they might not be exactly the size, shape, or colour you want. What if you could make any object you wanted, whenever you wanted? That is what a 3D printer does. It works like an ordinary computer printer except that, instead of using ink, it sprays soft plastic onto a board to build up a 3D object.

## ENERGY FOREVER

In 30 years time, the world could be using 50 per cent more energy. Where will we find it? Nuclear fusion (smashing atoms together) could be one way of making clean energy without harming the planet. It is the process by which energy is created inside the Sun. If we could build safe nuclear fusion plants on Earth, we could have clean energy forever.

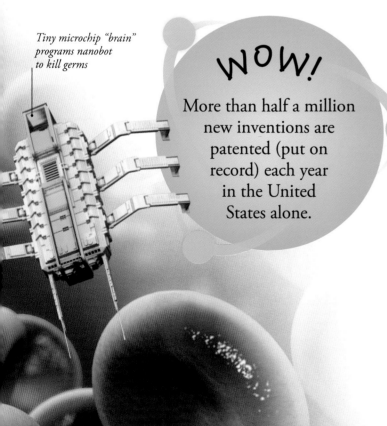

*Tiny microchip "brain" programs nanobot to kill germs*

**WOW!**

More than half a million new inventions are patented (put on record) each year in the United States alone.

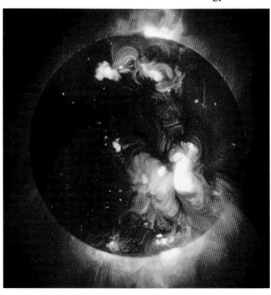

▲ SUN'S ENERGY *Nuclear fusion inside the Sun releases vast quantities of energy into space.*

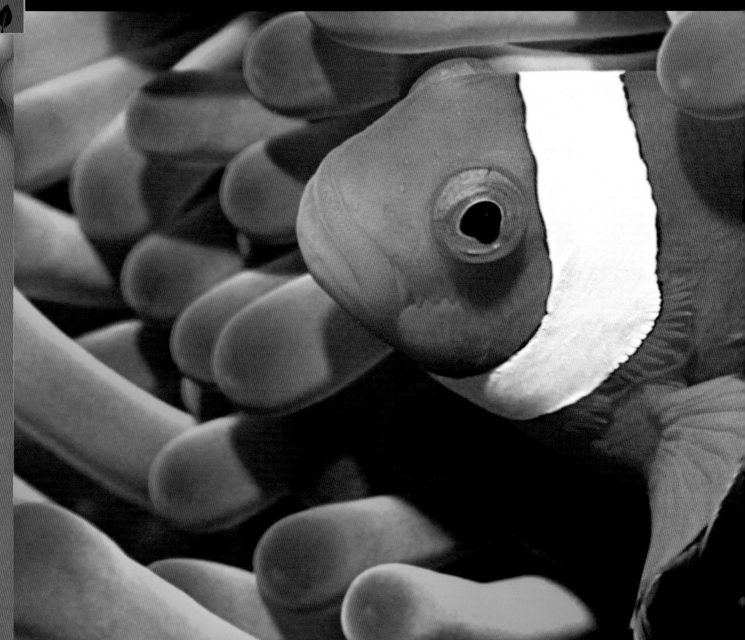

# LIVING ORGANISMS

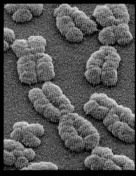

Our planet is full of living things, from tiny germs to giant trees, deep-sea fish to high-flying birds. Biology is the science of how living things work.

## LIVING PARTNERS

Sometimes two different species of living creatures form a partnership, known as symbiosis. This poisonous anemone protects the clownfish from attackers. In return, the fish keeps the anemone clean.

# Life on Earth

Almost every corner of planet Earth has life. Even tall mountains and frozen polar seas are home to amazingly tough animals and simple plants. But living things flourish best where it is warm and damp or wet – so tropical rainforests and coral reefs are filled with a dazzling array of life.

## LIFE ALL AROUND

The world around us teems with life. Birds and insects fly through the air, plants spring up from the soil, and reptiles and mammals climb the trees and crawl on or under the ground. Even seemingly "dead" matter is often packed with life. Water and soil are filled with vast numbers of microscopic life forms, too small for us to see.

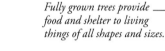
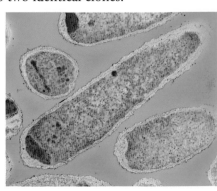

*Fully grown trees provide food and shelter to living things of all shapes and sizes.*

*Even bare soil is full of bacteria and fungi, and nourishes seeds that grow into plants.*

*Plants provide foods for some animals, which in turn are eaten by predators.*

▶ LIVING WORLD *From lowlands to high tree tops, the natural world is packed full of living creatures.*

## EARLIEST LIFE

Fossils are the remains of once-living things preserved in rocks. They show there were small, simple forms of life in the sea more than 3,000 million years ago. By 400 million years ago, plants and then animals appeared on land.

▶ FOSSIL IMPRESSION *Fossil imprints of strange soft-bodied creatures in the rocks of Ediacara Hills, South Australia, date back more than 550 million years. Many have no living relatives.*

## SIMPLEST LIFE FORMS

The simplest life forms are known as prokaryotes. They are made up of only a single, tiny cell, which can break down chemicals around it to obtain energy. To breed, they usually split into two identical clones.

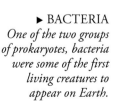

▶ BACTERIA *One of the two groups of prokaryotes, bacteria were some of the first living creatures to appear on Earth.*

## LONGEST LIVING

Giant tortoises famously reach great ages of 150 years old or more. But far older animals are ocean quahogs – clam-like shellfish that live for more than 500 years. Even these are young compared to the oldest of all life forms – bristlecone pine trees, which can live for 5,000 years.

▲ LONG-LASTING *Some giant tortoises have been measured at 180 years of age. Others are claimed to be more than 220 years old.*

▲ FOREST GIANTS *These giant redwoods are the heaviest living things, but their relatives, the coast redwoods, are taller, at up to 115 m (377 ft).*

## BIGGER THAN ALL

At 85 m (279 ft) high and 2,000 tonnes (2,204 tons) in weight, giant redwood trees are the biggest living things on Earth. Among animals, the blue whale is the largest – larger than even the great dinosaurs of long ago. The blue whale is 30 m (98 ft) long and weighs almost 200 tonnes (220 tons).

## WOW!

Tardigrades are microorganisms, about 0.5 mm (0.02 in) long. They can survive extreme conditions – even outer space.

## IN LARGE NUMBERS

Apart from worms and bugs such as flies, one of the most numerous wild animals is the red-billed quelea of Africa. However, among domesticated animals, chickens beat them, with a world population of over 20 billion – that is three chickens for every human.

◀ FLYING FLOCK
*There are about 1.5 billion red-billed queleas in the world. They are called "feathered locusts" because they eat so many farm crops.*

## IS IT EVEN ALIVE?

The lithops plant has leaves that look just like pebbles. Their shape, size, and colour makes them resemble small stones in their natural surroundings. They hardly grow or change at all, sometimes for years, earning the name "living stones". Then suddenly the plant produces new leaves or flowers.

▶ LIVING STONES
*Lithops blend in with stones as a means of protection, so animals rarely notice them to eat.*

# Classifying life

There are millions of different types of living thing on Earth. To understand how they relate to each other, scientists arrange them into groups. This is called "classification". Some groups are easy to identify – for example, all mammals feed their young on milk – but others are more complicated.

## GROUPING ORGANISMS

Scientists sort all living things into species. Organisms that can breed with each other are said to be the same species. Lions are a species, and so are humans. Species are grouped into larger and larger groups called families, orders, and classes. The biggest groups are called kingdoms, which are separated by how the organisms within them obtain the energy they need to live.

## MICRO LIFE

Some living things are so tiny, they can be seen only through a microscope. Each one is often made of a single unit, or cell. These single-celled life forms are neither plants nor animals. Some gain their energy from sunlight, as plants do, while others "eat" other organisms.

Paramecium are single-celled organisms found in freshwater.

## FUNGI

Most fungi feed on dead or dying plants and animals, by making them rot or decay. They then soak up nutrients and energy from the rotten food by absorbing them into their cells. We mostly see fungi when they grow as moulds, mushrooms, and toadstools.

Toadstools are fungi with a round cap on a short stem.

## PLANT

A plant gets its energy from the Sun. Substances in its body, usually in the green leaves, catch the energy in sunlight and use it to make sugars. The plant also takes in water and nutrients from the soil, and uses the sugar energy to grow.

African daisies are colourful flowers that flourish in sunlight.

## INVERTEBRATE

Animals get their energy by eating food, usually plants or other animals. Invertebrates are animals without backbones, such as worms, snails, insects, spiders, crabs, and starfish. They are mostly small but incredibly numerous.

Ladybirds are a type of insect, the biggest group of invertebrates.

## VERTEBRATE

Animals with inner skeletons and backbones are called vertebrates. Fish, amphibians, reptiles, birds, and mammals (including humans) are all vertebrates. They are usually larger than invertebrates, but exist in smaller numbers.

Snakes, such as this northern cat-eyed snake, are reptiles.

## FINDING RELATIVES

Humans have been working to classify living things for thousands of years. The earliest classifications were based on recognizing similar features in plants and animals, such as the way they feed their young, or the way their limbs are formed. Today, our understanding of genetics and evolution allows us to look at the DNA of living creatures to work out which ones have developed from the same kinds of ancestor.

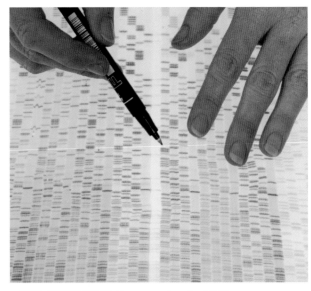

▲ RELATIVE DNA *By comparing the DNA of different species, scientists can look for links to creatures that lived long ago. This allows them to work out how different things evolved, and how they are related.*

▲ SIMILAR SHAPES *The rhinoceros (left) and the tapir (right) have similar shaped bones in their feet. This and other similarities show they are part of the same order (called "odd-toed ungulates").*

*Stubby, leg-like appendages*

## HARD TO CLASSIFY

Nearly all worms lack legs. But velvet worms have many pairs of bendy, stubby legs, as well as a bendy body like a worm or caterpillar. These creatures live in tropical forests and catch prey, such as insects, by squirting them with slime. Experts are not sure which other animals are their closest relatives. They are classified in their own group, Onychophora.

► VELVET WORM *These creatures creep across the floors of tropical forests, where they prey on insects.*

## RULE BREAKERS

Some living things have evolved very differently from others in the same group, sometimes taking up a different way of life. Plants such as the Venus flytrap, sundew, and pitcher plant get most of their energy from light, but they also "eat" small creatures by trapping and dissolving their bodies. The tiny animals that make coral reefs can absorb energy from sunlight, as plants do, by partnering with microorganisms living inside them.

▲ VENUS FLYTRAP *Unlike most plants, this species can "eat" insects. Its leaves snap shut to trap wandering flies inside.*

▲ CORAL POLYPS *These polyps manage to gain energy from sunlight, as plants do, thanks to microorganisms living inside them.*

# DNA and genes

Animals start life as a single tiny cell called a zygote that slowly grows into a full-sized body. The instructions for how this happens are contained in the form of chemicals called deoxyribonucleic acid (DNA). Adults produce young, and pass their DNA on to them. This means offspring look a bit like their parents.

## WHAT IS DNA?

DNA is a long, thin molecule that can be found in almost every cell in your body. Each length resembles a ladder twisted like a corkscrew, known as a double helix. DNA can copy itself by dividing into halves. Each half then copies itself, resulting in two lengths of DNA that are the same as the original. This is how DNA is copied in parents, to pass on to their offspring.

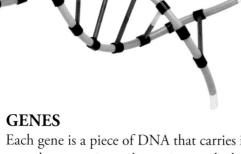

◄ TWISTED LADDER
*To divide in half, DNA splits lengthways – midway across each "rung" of the ladder.*

## GENES

Each gene is a piece of DNA that carries information to make a protein, and proteins are the building blocks of bodies. Together, our genes decide the size, shape, and colour of our bodies, as well as how each organ works, such as a muscle that pulls, an eye that detects light, and a stomach that digests food.

▲ GENES FOR EVERY BODY PART *Cats' genes have instructions to make all parts including eyes, ears, claws, and different colours of fur.*

## CHROMOSOMES

Information in DNA is carried by the order of the various chemical parts along the "rungs" of its ladder. In each living thing the DNA is in separate lengths called chromosomes. A human body has two sets of 23 chromosomes, one set from each parent. A pigeon has two sets of 40, and a mosquito has two sets of three.

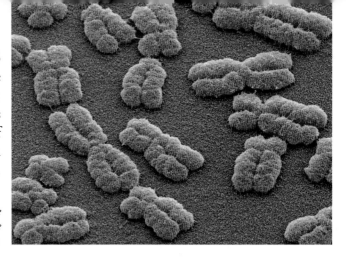

▶ X-SHAPE *When the length of DNA in a chromosome folds and coils up, it forms an X-shape. These are human chromosomes seen under a microscope.*

## SAME BUT DIFFERENT

The way that genes are passed on means that each offspring receives a slightly different combination of genes compared to its parents and other offspring. So sisters and brothers resemble each other and their parents, but each individual has slightly different features. That is, except for identical twins, who have exactly the same genes as each other.

### HOW MANY GENES

Different species of living things have different numbers of genes. However, bigger ones do not always have more. Some tiny plants and creatures have huge numbers, while some big animals and trees have fewer. A human has around 20,000 genes.

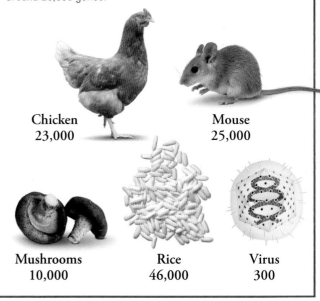

Chicken
23,000

Mouse
25,000

Mushrooms
10,000

Rice
46,000

Virus
300

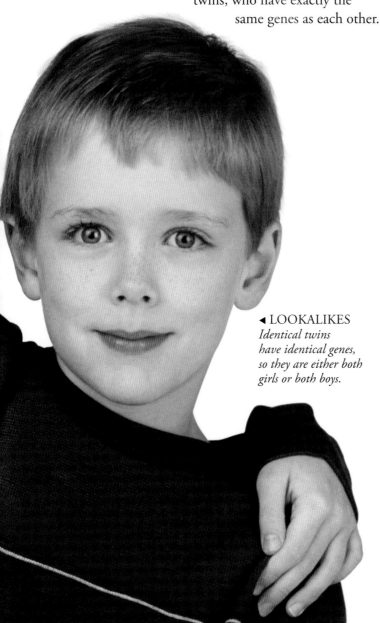

◀ LOOKALIKES
*Identical twins have identical genes, so they are either both girls or both boys.*

## UNUSUAL GENES

Sometimes genes change because DNA does not copy itself exactly. The changed gene may make an animal or plant look and work differently. For example, if the genes for the colour of skin, fur, or feathers do not work properly, the result is a mainly white animal called an albino.

▶ WHITE BIRD
*This albino hummingbird has unusual genes, giving it white feathers but pink legs and a pink beak.*

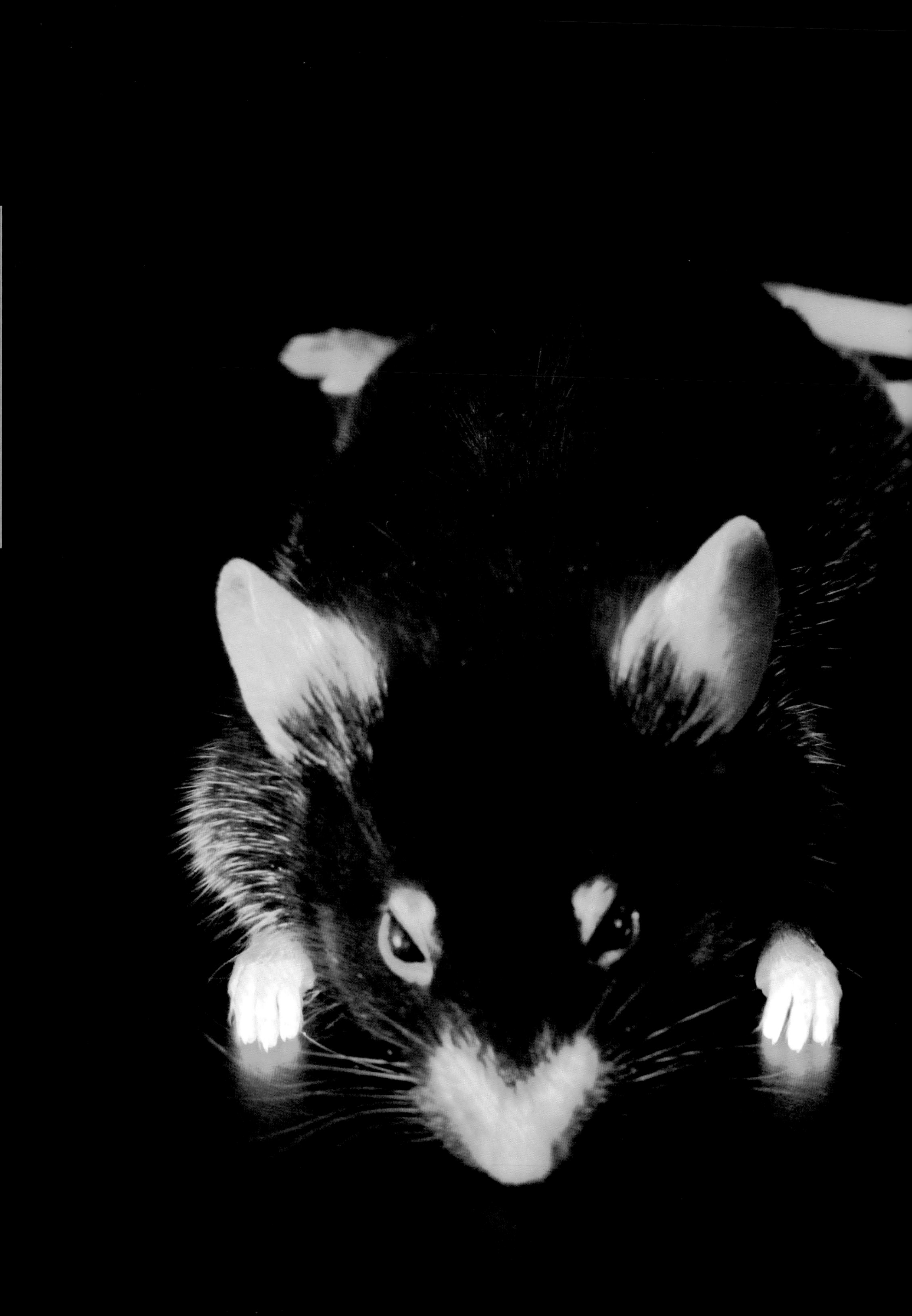

**THE POWER OF GENETICS**
By adding genes from one organism to the DNA of another, we can change the chemicals produced inside its body. This mouse has jellyfish genes added to its DNA, so its skin glows in the dark. Other experiments have created crops that are resistant to disease, or which produce tastier fruit.

# Evolution

Species of animals and plants change because of genetic variations that occur with each group of offspring. Organisms born with more advantageous features will be more likely to better survive. These features are passed on to future generations in a process known as natural selection. Over millions of years this has allowed primitive life forms to develop into complex animals and plants.

## ADAPTATION

Each animal or plant has features or adaptations to help it stay alive. Desert creatures cope with extreme heat and little water, while polar animals have extra fur, feathers, or fat to stay warm. This also means that most living things can survive only in their own habitat, so parrots cannot live on icebergs, and penguins cannot live in tropical forests.

**WOW!**

Out of every 1,000 species that have ever lived on Earth, 999 have become extinct.

▶ ADAPTED TO THE DESERT *The Bactrian camel has many adaptations to cope with the lack of water, great summer heat, and severe winter cold in its Asian desert home.*

*Hump of fat stores energy*

*Long eyelashes keep out sand*

*Thick fur keeps out cold in winter*

*Wide hooves spread weight on soft sand or snow*

# EXTINCTION

Over thousands or millions of years, as new species appear or evolve, others gradually die out, or become extinct. This happens because some species cannot evolve fast enough and alter their adaptations to cope with the changing environment and new species around them. This has happened for billions of years, ever since life first appeared on Earth.

▲ WOOLLY MAMMOTH *This mammoth evolved to survive Ice Age conditions, but after that it could not survive the warming climate or protect itself from being hunted.*

# SURVIVAL TACTICS

Living things have many kinds of adaptations that help them survive, such as big teeth and claws in predators, and long legs in prey for running away fast. Many creatures are camouflaged – that is, they are coloured so that they blend into their surroundings. This makes them harder to see, so they can hide from predators or sneak up on prey.

▲ GREEN ON GREEN *The green iguana lizard is the same colour as the leaves of its forest home. This camouflage works best if it stays still when enemies are near.*

# EVOLVING TOGETHER

Living creatures evolve (adapt) to cope with their surroundings, and with other organisms living alongside them. Sometimes, very different organisms can look quite similar because they have adapted to live in the same environment – for example, whales and dolphins are mammals, but their bodies are shaped like fish to help them move through the water easily. Some species have evolved to live in partnership with each other, while others have taken on the colours of dangerous creatures so that predators avoid them.

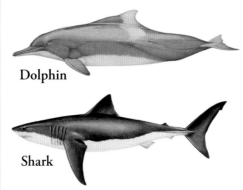

Dolphin

Shark

◄ LOOK SIMILAR *The dolphin is a mammal and the shark is a fish. But they have evolved to have similar streamlined body shapes to swim at speed in water. This is known as convergent evolution.*

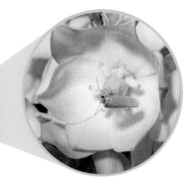

◄ INTERDEPENDENCE *Yucca moth caterpillars eat seeds of the yucca plant. In return, the moth carries yucca pollen so the seeds can develop. This is called co-evolution.*

Coral snake

Milk snake

▲ LOOK DANGEROUS *The milk snake has no venom but it has evolved to look like the venomous coral snake. This is called mimicry.*

# Simple life forms

All living things are made of tiny parts called cells, which are so small they can only be seen through a microscope. The human body has more than 100 million million cells (see p.260). But many forms of life are just one cell each. Some get energy from light (as plants do). Others are like animals and eat even smaller things. Still others do both.

## BACTERIA

Bacteria are simple cells with few parts inside. They are so tiny that about 100,000 bacteria would fit in this "o". Bacteria live in their trillions almost everywhere – in soil and water, on rocks and walls, and floors, under icebergs, in boiling-hot springs, and on and inside living things. Some bacteria cause illness and make up one of the groups of harmful life forms we call germs. Some bacteria are helpful to humans and can protect us from illnesses. Others are used to make food such as yoghurt and cheese.

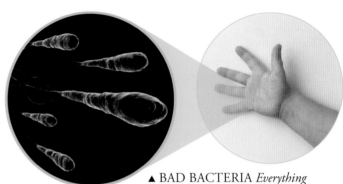

▲ BAD BACTERIA *Everything we touch has many kinds of bacteria. Washing hands helps to get rid of them so the bad bacteria do not get inside us.*

## MANY DIFFERENT TYPES

There are more than 100 main groups of single-celled living things. Diatoms are like mini-plants that live in any water, from damp gutters to the open ocean. Forams mostly live in the sea and catch bits of food using long tentacles. Flagellates swim along by waving long, thin parts known as flagella.

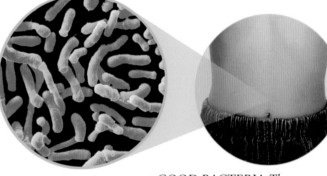

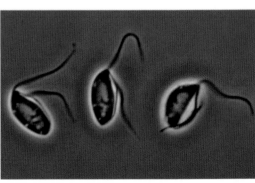

▲ GOOD BACTERIA *There are millions of helpful bacteria inside our guts. They can destroy harmful bacteria and help digest our food.*

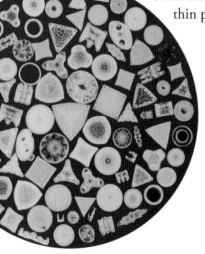

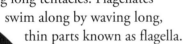

▲ DIATOMS *These make cases around themselves of silica, the mineral of sand, in amazing shapes such as balls, pyramids, and boxes.*

▲ FORAMS *A foraminiferan builds a shell, or test, with small holes for its tentacles. The test lasts after the foram inside dies.*

▲ FLAGELLATES *Some flagellates collect energy from sunlight, others feed on tiny prey. They thrash their flagella to and fro like long whips to move along.*

## INSIDE THE CELL

Single-celled organisms can be divided into two different types based on their cell structures. Some, such as bacteria, have just a few separate parts inside the cell, and no blob-shaped nucleus or control centre. These organisms belong to the group of single-celled organisms known as prokaryotes. Other single-celled creatures, called eukaryotes, do have a nucleus.

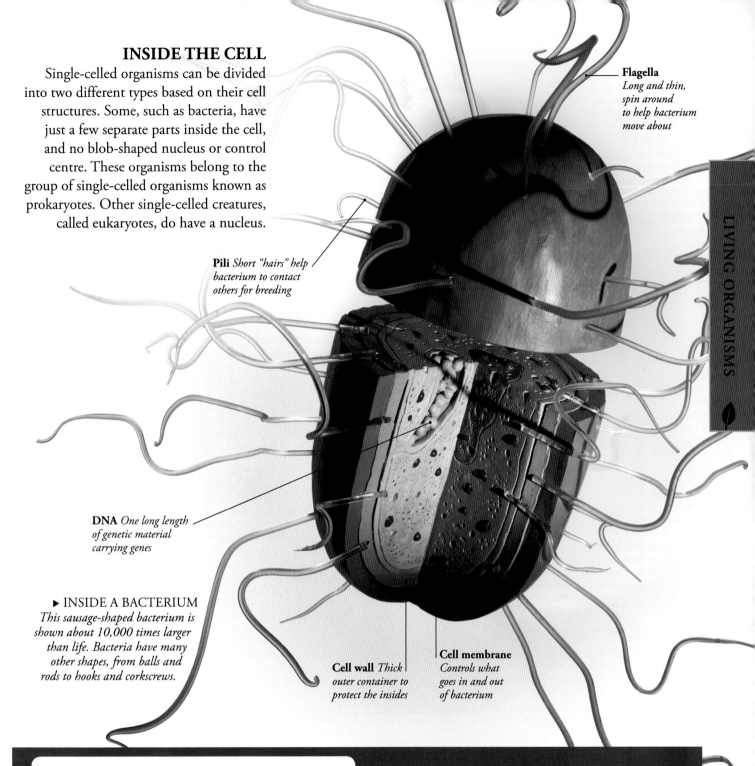

**Flagella** *Long and thin, spin around to help bacterium move about*

**Pili** *Short "hairs" help bacterium to contact others for breeding*

**DNA** *One long length of genetic material carrying genes*

▶ INSIDE A BACTERIUM
*This sausage-shaped bacterium is shown about 10,000 times larger than life. Bacteria have many other shapes, from balls and rods to hooks and corkscrews.*

**Cell wall** *Thick outer container to protect the insides*

**Cell membrane** *Controls what goes in and out of bacterium*

## GERMS AND DISEASE

More than 2,000 kinds of single-celled living things produce illness in humans. Bites of certain mosquitoes spread a microbe called plasmodium that causes the serious fever malaria. Bacteria in bad food or dirty water multiply in the gut to cause vomiting and diarrhoea, as in cholera or typhoid. Single-celled fungi cause itchy redness in nail infection and athlete's foot. Viruses spread by sneezes and coughs produce colds, flu, and measles.

**Plasmodium: Malaria**

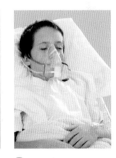
**Bacteria: Cholera**

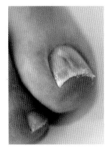
**Fungus: Nail infection**

**Virus: Cold, flu**

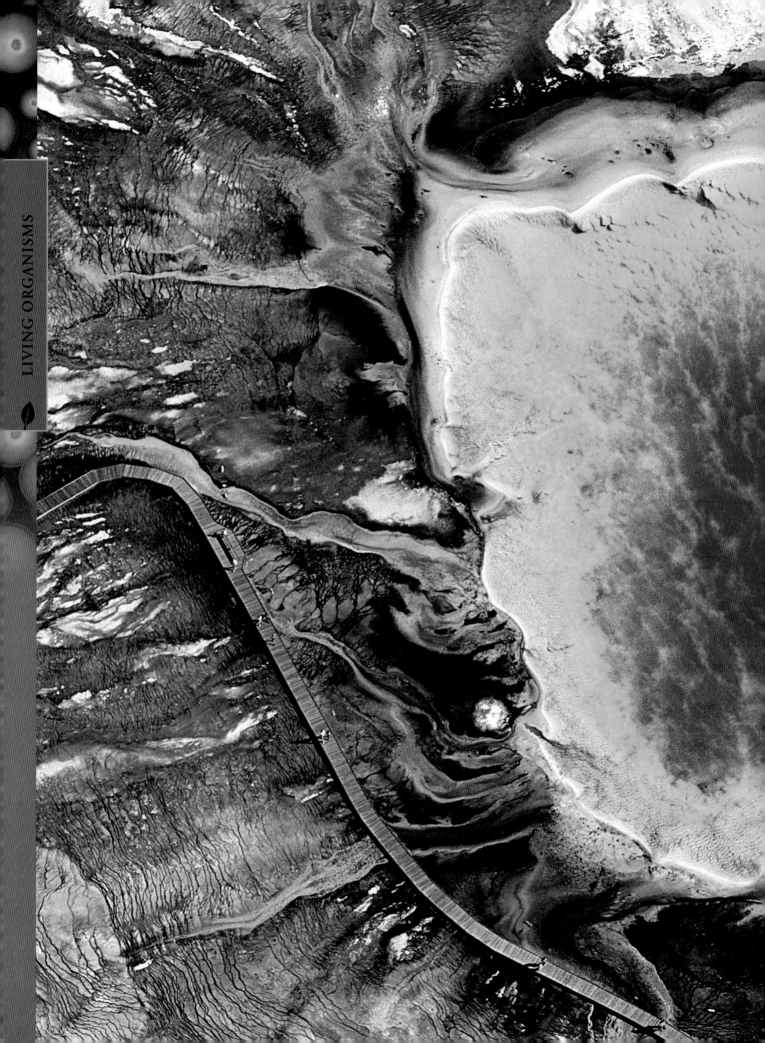

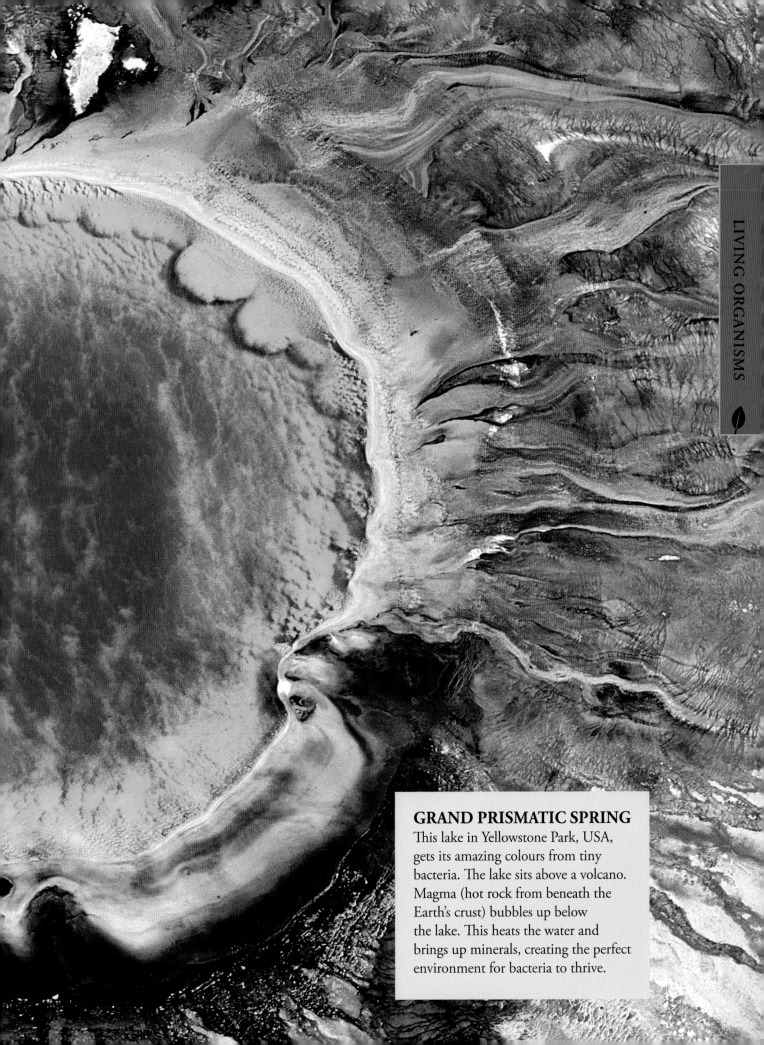

### GRAND PRISMATIC SPRING
This lake in Yellowstone Park, USA, gets its amazing colours from tiny bacteria. The lake sits above a volcano. Magma (hot rock from beneath the Earth's crust) bubbles up below the lake. This heats the water and brings up minerals, creating the perfect environment for bacteria to thrive.

# Fungi

A fungus gets its energy and nutrients from other living things – or more often, dead ones. The fungus makes substances called enzymes that ooze out of it, into dead bodies of plants or animals. These enzymes break down the body into nutrients that the fungus can absorb for food.

## FUNGAL NETWORK

A fungus feeds using long, thread-like parts called hyphae. These grow through the soil, into and around dead plants and animals of all kinds. The hyphae form a network called a mycelium. They cause decay and rotting, and take in the nutrients.

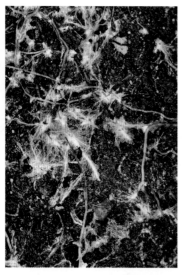

◄ WHITE THREADS *These pale threads, hyphae, form a web-like net in the soil.*

## FUNGAL SPORES

A fungus breeds by making tiny seed-like spores. To do this the fungus grows upwards through the soil and produces a part above the surface, which is the fruiting body – usually called a mushroom or toadstool. This makes millions of spores that float away in the wind to grow elsewhere.

◄ PUFFING SPORES *The puffball is a ball-shaped fruiting body that blows out tiny spores, which look like dust.*

## TASTY AND DEADLY

There are many different kinds of mushroom and toadstool. Some are tasty and safe to eat. But others are poisonous and can cause serious illness, sometimes even death. There is no simple way to see the difference, and only experts can tell which wild mushrooms are safe and which are dangerous.

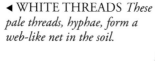

◄ DELICIOUS *Fungi such as button mushrooms are sold in markets. They are safe and tasty to eat.*

## USEFUL FUNGI

Fungi grow on rotten food, which can be dangerous to us if we eat it. But they also have many uses. In nature, they help to turn plant and animal remains back into useful soil. We also use a kind of fungus called yeast to make fresh bread and to brew beer.

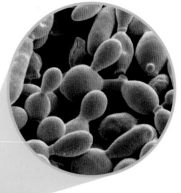

▲ BUILD-UP *Yeasts are microscopic fungi. They produce gas bubbles that make bread rise.*

◄ BREAKDOWN *When fungi grow on something, they begin to break it down. We say it has gone mouldy or rotten.*

### COMPOST HEAP

Fungi are nature's recyclers. In a compost heap, many kinds of fungi rot down bits of plants, for example leaves, shoots, flowers, cuttings, and leftover food such as vegetable peelings. After weeks or months the result is brown, crumbly, and nutrient-rich, and ready to be added back into the soil.

Well-rotted compost

## BRACKET FUNGI

Some fungi grow on the sides of trees and other wood, forming round ledges like curved shelves. These are bracket fungi. They send their string-like hyphae into the tree's wood to get nutrients. After a time, bracket fungi may cause the tree to sicken and die. Some brackets grow so large and hard that people can stand on them.

◄ DANGEROUS *The fly agaric, recognized by its white-spotted red cap, is poisonous. If eaten, it causes sickness, twitching, and sleepiness.*

## WOW!

The football-sized giant puffball fungus produces more than a million million spores from one fruiting body.

▲ SCARLET BRACKET *This bracket fungus from Australia and New Zealand ranges from brown to red-pink. It grows on living trees and dead stumps.*

229

# Plants

Like other major groups of living things, plants are usually identified by how they get their energy – by using sunlight to build up complex substances from simple ones (see p.232). Most plants are green, but some have red or yellow leaves and other parts.

## TYPES OF PLANT

Most plants are made of the same vital parts – roots, stems, and leaves. The main subgroups of plants are divided by the way they breed or reproduce, and by which special parts they have.

| | | |
|---|---|---|
| **ALGAE AND SEAWEEDS** | Some experts do not classify these as true plants. They range from tiny – just a few cells – to giant seaweeds as big as trees. |  |
| **MOSSES** | Most mosses are small, with simple leaves on short stems, and no proper roots. They breed by spores (see p.234). |  |
| **FERNS** | A typical fern has tubes in its stems which end in leaflike fronds, and roots to take in water and nutrients, but no flowers. |  |
| **CONIFERS** | Nearly all conifers have needlelike leaves that they keep all year round. They make their seeds in woody cones. |  |
| **FLOWERING PLANTS AND TREES** | The biggest plant group, these are the most complex plants with flowers that produce seeds, often in fruits. |  |

## ABUNDANCE OF PLANTS

The main feature of each kind of living place, or habitat, is its plants. Oak trees dominate an oak wood, pine trees rule a pine forest, and grasses cover the savanna and prairie. The kinds of plants, in turn, determine which animals live there.

▼ BURSTING WITH GREEN
*Warm and damp all year, tropical rainforests have many different kinds of plants in a small area.*

## PLANT CELLS

Under the microscope, plant cells differ from animal cells. A typical plant cell has a thick, stiff outer covering called the cell wall. It also has bag-like, liquid-filled parts inside known as vacuoles, and several tiny chloroplasts, which make the plants green and help them produce food.

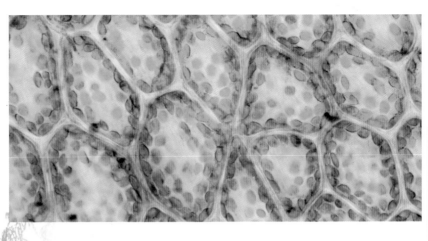

▶ WALLED PATTERN *Leaf cells are separated by thick cell walls. Each leaf cell has many small green blobs called chloroplasts.*

## WOW!

The total weight of all the Earth's plants is 1,000 times more than the total weight of all the animals.

## PLANTS AS FOOD

Nearly all animal life depends on plants, directly or indirectly – herbivores eat the plants, carnivores eat the herbivores, and scavengers feed on dying and dead plant matter. Humans rely on plants too, grown in farms, orchards, and greenhouses.

◀ TALLEST HERBIVORE
*The giraffe can reach up to leaves 7 m (23 ft) high – higher than any other ground-based herbivore.*

## PLANT DEFENCES

Plants have many ways to defend themselves against herbivorous animals. They develop horrible-tasting or poisonous substances, stinging hairs, sharp thorns and spines, and hard casings, as in nuts. Animals soon learn from their parents or their own experience to avoid these types of plants. Other plants simply grow very fast to keep up with being eaten.

▲ POISONS *The red colour of the poisonous baneberries is a warning that animals learn to avoid.*

▲ STINGS *The stinging nettle plant has small hairs on the leaves and stem that break off to release an irritant fluid.*

▲ SPINES *In a cactus, the leaves are modified to become sharp spines that protect the green stem.*

# Photosynthesis

Plants make their own food through a process called photosynthesis, which means "making with light". A plant catches the energy in light rays and converts it into chemical energy in sugars. The plant then uses these high-energy sugars to live, develop, and create flowers and seeds that will grow into new plants.

## IN AND OUT

In photosynthesis, the energy from sunlight is used to join carbon dioxide ($CO_2$) from the air with water taken in by the plant, usually through its roots from the soil. The result is sugars, or carbohydrates, which spread around the plant dissolved in a fluid, called sap. This process also releases oxygen into the air.

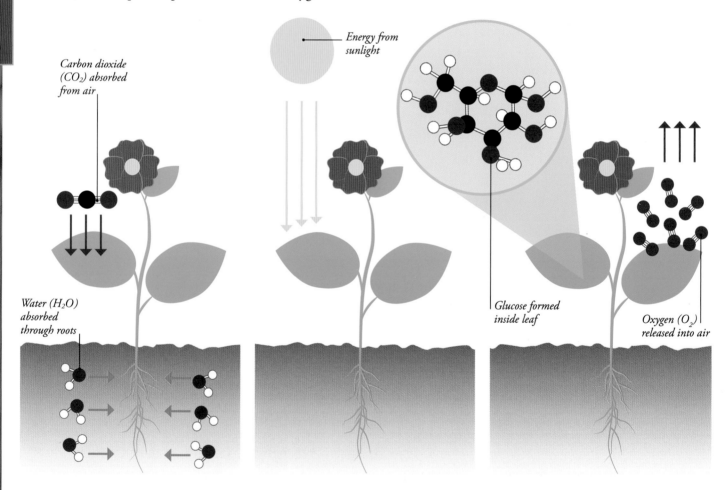

Energy from sunlight

Carbon dioxide ($CO_2$) absorbed from air

Glucose formed inside leaf

Oxygen ($O_2$) released into air

Water ($H_2O$) absorbed through roots

$$H_2O + CO_2 \quad + \quad Sunlight \quad = \quad Sugars + O_2$$

▲ WATER AND CARBON DIOXIDE
*Plants take in carbon dioxide ($CO_2$) from the air, and water ($H_2O$) from the soil.*

▲ SUNLIGHT *Light, a form of energy, is taken in by leaves. Mainly red and blue colours are absorbed, leaving green to be reflected and give leaves their colour.*

▲ GLUCOSE AND OXYGEN *The reaction produces a simple sugar called glucose, and releases oxygen.*

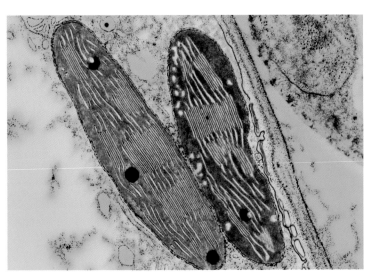

## WHERE PHOTOSYNTHESIS HAPPENS

The microscopic cells of a leaf have dozens, sometimes hundreds, of flat disc- or sausage-shaped parts inside known as chloroplasts. These have thin layers of a green pigment called chlorophyll. This pigment helps the plant's leaves trap energy from the Sun.

◀ CHLOROPLASTS *Each chloroplast is less than 0.01 mm long, with many thin layers inside. A typical leaf cell contains up to 100 chloroplasts.*

## WOW!

Tropical rainforests produce nearly 40 per cent of the world's oxygen through photosynthesis.

## LIGHT AND DARK

Plants need light to grow. If a plant does not have enough light, it may grow fast for a time, trying to get above anything nearby that causes shade. But if light is still lacking, eventually it dies.

▶ NOT ENOUGH *Compared to the plants grown in normal light on the left, the plants on the right have grown in too little light. They are pale and try to grow too fast.*

## NOT ALL GREEN

The main plant coloured pigment, chlorophyll, is green. But a range of other plant pigments also carry out photosynthesis. They include orange or yellow carotenoids (as in carrots), red or yellow betalains, and red, purple, or blue anthocyanins.

▶ PIGMENT FOR COLOUR
*The flowering plant* Solenostemon *"Black Prince", has dark purple leaves because of pigments called anthocyanins.*

### NO PHOTOSYNTHESIS

Some plants are not green and do not carry out photosynthesis. They are parasites, stealing nourishment from others. This almost-white plant, called Indian pipe, is a parasite on fungi and trees. Indian pipe is usually found on the remains of dead plants.

**Parasitic Indian pipe**

# Plant life cycles

Plants have many different ways of breeding. Some simply grow new, smaller plants from parts of their bodies. Others break off leaves, stems, or shoots that grow into new plants. Simple plants, such as mosses and ferns, make tiny dust-like particles – spores – that grow into new plants. Flowering plants develop flowers which make seeds, which are contained in fruits.

## ONE YEAR OR MANY?

Some plants have a one-year life cycle. They produce flowers that develop seeds. The adult plant dies but next year the seed grows into a new adult plant and the cycle begins again. Other plants, such as trees, live for hundreds of years.

▶ ANNUAL CYCLE *Plants that have a generation every year, such as poppies, are called annuals. Those which survive many years are called perennials.*

*2. Petals wither and die, leaving a flask-like capsule or pod containing seeds.*

*3. Poppy seeds are tiny specks that spread easily in the wind or are carried by animals.*

*4. Next spring the seeds grow, or germinate, into new poppy plants.*

*1. Poppy plants grow quickly in summer and develop their flowers.*

## FLOWERS

The flower, or bloom, is a plant's breeding part. A typical flower has several fan-like petals around a central set of male and female parts. The male parts make pollen, the female ones have unripe seeds. Pollen must reach the female parts so the seeds can ripen.

*A flower's bright colours and sweet smell attract insects.*

*Pollen sticks to insects and birds and is carried to new flowers.*

◀ POLLEN GRAIN *Bees carry pollen in sacs on their legs. Pollen looks like dust to the naked eye, but under a microscope, it has different shapes such as parachutes, balls, or baskets.*

## THE FOUR SEASONS

Many perennial plants change with the seasons of the year, especially bushes, shrubs, and trees. A deciduous tree (one that sheds leaves annually) blossoms, or flowers, in early spring. Fresh new leaves appear in late spring and early summer. It then starts losing its leaves, usually in autumn, when they turn brown, wither, and fall. In winter the branches are bare.

**Tree blossoms in spring**

**Leaves appear in summer**

**Browned leaves fall in autumn**

**Tree is left bare in winter**

## SEEDS AND SPORES

Some plants, such as mosses, liverworts, ferns, and horsetails, grow from spores rather than seeds. Unlike a seed, a spore has no stores of food for its early growth, or germination. Spores are usually very tiny and blow easily in the wind, then germinate if they find a suitable patch of bare, nourishing soil.

▶ FERN BUTTONS *Ferns, such as this goldfoot fern, grow spores in brown "buttons", called sori, on the underside of the fronds (leaves).*

## GROWING WITHOUT SEEDS

Some plants can create exact copies of themselves without using seeds or fruit. Instead they send out a special shoot called a runner, or stolon. A new plant grows at the end of the runner, and takes root in the nearest available soil.

**WOW!**

The biggest seed is from the coco-de-mer palm tree. It is larger than a football.

*Parent plant produces runners*

*Runners take root in nearby soil*

▲ NATURAL CLONES *Strawberry plants breed using runners. The new plants that grow are clones, that is, they have the same genes as the original.*

*New plants can send out their own runners to grow more clones*

235

# Invertebrates

Invertebrates are animals that lack a backbone, or spine. Most invertebrates are small but very common, thriving in almost every habitat in the world. But a few are huge – the biggest is the deep-sea colossal squid, growing to 14 m (46 ft) long and weighing half a tonne.

## NO BACKBONE

Insects are the biggest group of invertebrates. An adult insect has six legs, usually two or four wings, and a body divided into three main parts – head, thorax, and abdomen. An outer skeleton, called the exoskeleton, covers and protects its entire body. Within the insects, beetles are the largest subgroup – there are more than 300,000 known species of beetle.

Head has feelers or antennae, mainly used as an organ of smell

Front wings are hard covers for rear wings

Rear wings are large for flight

Abdomen contains digestive and other internal parts

Three pairs of legs joined to thorax

▲ SMALL AND STRONG *The Atlas beetle is one of the strongest animals on Earth, and can carry 100 times its own weight on its back.*

## TYPES OF INVERTEBRATE

This tree diagram shows the main groups of invertebrates and how they are related to each other. Sponges and jellyfish are mostly simple creatures without a brain or heart. Starfish are animals with a body that is radial – like a wheel with spokes (other creatures have left and right sides). The arthropods or "joint-legs" include many familiar invertebrates both on land and in water. They include insects, arachnids (spiders and scorpions), and crustaceans such as crabs.

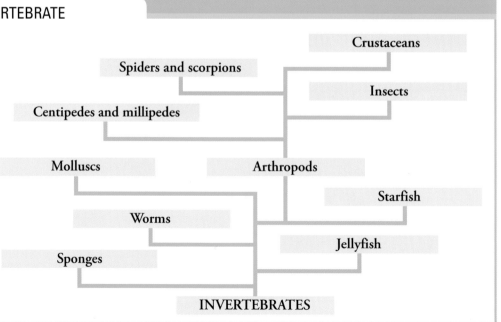

Crustaceans

Spiders and scorpions

Insects

Centipedes and millipedes

Molluscs

Arthropods

Starfish

Worms

Jellyfish

Sponges

INVERTEBRATES

## SPONGES

Sponges have a basic body structure with no defined organs – they filter water to get nutrients. Most are spongy, while some, such as glass sponges, have a hard supporting structure.

**Stove-pipe sponge**

## MOLLUSCS

Molluscs include shelled animals, such as clams, mussels, and limpets. Most live in the sea, but some, such as snails, live on land. The octopus is also a mollusc.

**Garden snail**

## JELLYFISH

Jellyfish, corals, and anemones are members of a group called cnidaria. They have special stinging organs called cnidocysts, which release venom when touched or triggered.

**Jellyfish**

## INSECTS

One of the most diverse animal groups, insects have six legs and two sensory antennae. Bugs, butterflies, bees, and beetles are all insects.

**Desert locust**

## STARFISH

Starfish, sea cucumbers, and sea urchins make up the group called echinoderms. Many of these can regenerate their tissues if damaged, for instance, some starfish can regrow lost arms.

**Sea star**

## SPIDERS AND SCORPIONS

Spiders and scorpions, part of the group called arachnids, have eight legs and sharp mouthparts. Some spiders spin elaborate webs to catch prey.

**Red-kneed tarantula**

## WORMS

There are many different groups of worms – flatworms, roundworms, and segmented worms. Each group of worms has thousands of different species within it.

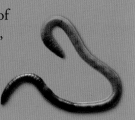

**Earthworm**

## CRUSTACEANS

Most crustaceans, including lobsters, barnacles, and crabs, live in water. They often have hard outer shells to protect their bodies.

**Lobster**

## CENTIPEDES AND MILLIPEDES

Centipedes and millipedes have long bodies divided into many segments. Centipedes have one pair of legs per segment and millipedes have two pairs per segment. Their bodies are covered in a tough but flexible exoskeleton.

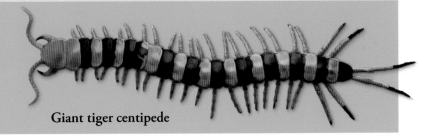

**Giant tiger centipede**

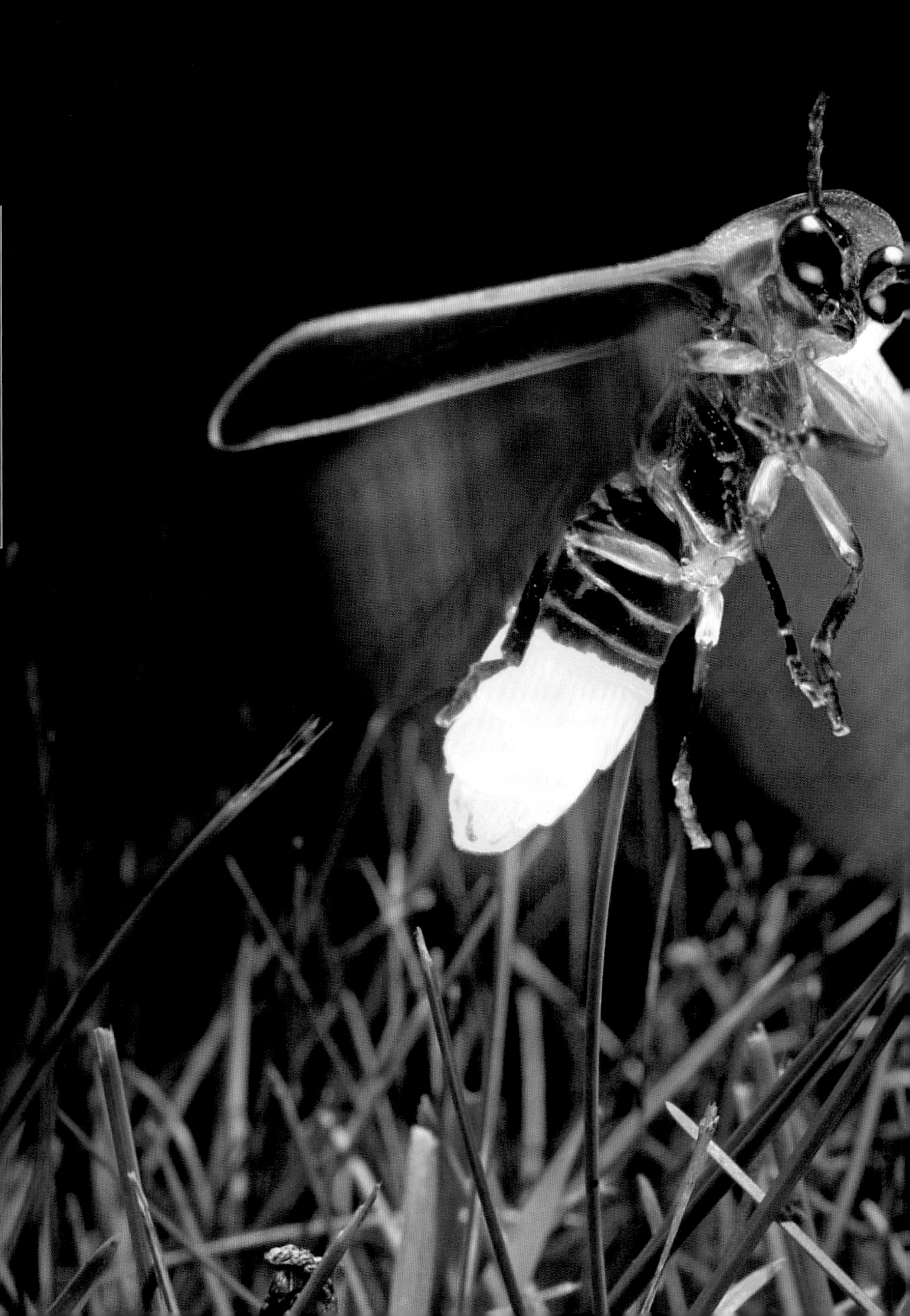

## FIREFLY

These amazing insects can produce light from their bodies. Special chemicals inside them create a yellow or green glow in a process called "bioluminescence". The bugs use this light to attract mates. They can use a steady glow or flashing lights to signal to each other.

# Vertebrates

Most of the world's large animals, from sharks to humans, are vertebrates. Their common feature is a strong framework inside the body, called a skeleton. This is based on a vertebral column – also known as a backbone, or spine.

## SKELETON AND VERTEBRAE

A backbone is a long row of smaller bones called vertebrae, joined like links in a chain. At the front end are the bones of the skull. In most vertebrates the bones at the rear end become smaller and narrower, forming the tail. Limbs such as fish fins, bird wings, and animal legs are attached to the vertebral column. The number of vertebral bones varies from fewer than 10 in frogs to more than 700 in the deep-sea snipe eel.

*Skull forms the shape of the head*

*Neck region has cervical vertebrae*

*Main body has thoracic and lumbar vertebrae*

*Hip region has sacral vertebrae*

*Tail has caudal vertebrae*

## WOW!

Only about 4 per cent of animal species are vertebrates. The rest are invertebrates.

▲ SUPPORT SYSTEM *The backbone, or vertebral column, usually runs along the middle of the body. This is the vertebral column of a giant salamander (right), the largest of the amphibian vertebrate group.*

## TYPES OF VERTEBRATE

This tree diagram shows how the main vertebrate groups are related to each other. The earliest vertebrates were jawless fish, more than 500 million years ago. Other fish groups then appeared. Some lobe-finned fish developed legs and became the first land vertebrates, called amphibians. From these came the reptiles, including dinosaurs. One dinosaur group developed into birds, another reptile group gave rise to mammals.

Mammals

Reptiles

Birds

Fish

Amphibians

**VERTEBRATES**

## FISH

Fish are cold-blooded – they cannot make their own body heat. They use fins to move, gills to breathe, and most are covered in scales. There are jawless fish such as lampreys, cartilaginous fish such as sharks and rays, and bony fish such as salmon and cod.

**Regal tang**

 Usually lay eggs to reproduce

 Live underwater

 Absorb oxygen from the water using gills

 Swim with the help of fins and a tail

 Most are cold-blooded

## AMPHIBIANS

Amphibians are cold-blooded and have moist skin. Most lay eggs in water. These hatch into larvae, called tadpoles, that change shape (metamorphose) as they grow into adults. Amphibians include those with tails, such as newts and salamanders, and those without tails, such as frogs and toads.

**Poison dart frog**

 Usually lay eggs to reproduce

 Have moist skin and may die if they dry out

 Many can survive in water and on land

 Some hatch as tadpoles and change shape to become adults

 Are cold-blooded

## REPTILES

Reptiles are cold-blooded. Most have scales for protection. Most lay eggs but some give birth to babies. Reptiles include turtles, terrapins, and tortoises; lizards and snakes; and crocodiles, alligators, and caimans.

**Savanna monitor lizard**

 Most lay eggs to reproduce

 Have dry, scaly skin

 Most are meat eaters

 Most live in warmer climates

 Are cold-blooded

## BIRDS

Birds are warm-blooded (making their own body heat). They have wings, feathers, and a beak, and most fly. Birds lay hard-shelled eggs and care for their baby chicks. There are more than 30 groups, from huge ostriches to tiny hummingbirds, fierce eagles, and all kinds of garden birds.

**Pigeon**

 Lay eggs to reproduce

 Have beaks instead of teeth

 Are covered in feathers

 Have wings and most can fly

 Are warm-blooded

## MAMMALS

Mammals are warm-blooded. They have hair or fur for warmth and protection. A few, such as echidna and platypus, lay eggs; the rest give birth to babies. The females produce milk to feed the young. The wide variety of mammals form more than 30 groups.

**Black panther**

Almost all give birth to live young

Feed their young on milk

Most have hair or fur

Are warm-blooded

Include human beings

LIVING ORGANISMS

241

# Energy from food

Different animals get their energy from different kinds of food. Some are fierce hunters with sharp teeth and pointed claws. Others munch plants for most of the day. Leeches and bugs such as fleas suck blood, while some beetle grubs chew solid wood and earthworms eat soil.

## HERBIVORES AND CARNIVORES

A herbivore is a creature that obtains all or most of its food from plants. Some herbivores, such as caterpillars, eat only one kind of plant, while others, such as elephants, eat shoots, flowers, fruits, and leaves of a wide variety of plants. A carnivore gets all or most of its food from eating other animals. Carnivores have special body parts, such as sharp teeth and claws, to catch and chew food.

*Wide back teeth and deep lower jaw for chewing*

▲ HERBIVORE ADAPTATIONS *A goat has wide-topped back teeth – premolars and molars – that crush and grind tough plants, to get as many nutrients as possible from them.*

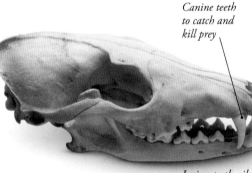

*Canine teeth to catch and kill prey*

*Incisor teeth nibble meat from bone*

▲ CARNIVORE ADAPTATIONS *The fox has long pointed front teeth, called canines, and sharp back teeth to slice up flesh and bone.*

◄ FOOD FOR YOUNG *Young carnivores may not be able to catch prey for themselves. Their parents bring them a share of their own meals.*

## OMNIVORES

An omnivore eats a mixture of plants and animals. Many omnivores change their diet with the seasons, depending on what is available. For example, bears mostly live on fruits and berries, but will eat meat when it is available.

▲ WIDE DIET *The raccoon eats all kinds of foods, from bugs, mice, and eggs to seeds, flowers, fruits, and human leftovers, so it can live almost everywhere.*

## PARASITES

A living thing that gets food or shelter from another – its host – and harms this host in the process, is called a parasite. Some parasites suck blood or other body fluids from outside. Others, such as tapeworms and flukes, live inside the host's body.

▲ BLOODSUCKER *Fleas infest mammals and birds, sucking blood through their sharp, piercing mouthparts.*

## DETRITIVORES

Detritivores feed on detritus – general bits and pieces of dead animals and plants, such as old leaves, decaying fruit, animal droppings, and rotting eggs. They help to recycle nutrients that other creatures waste.

▲ DEAD AND DECAYED *Woodlice feed mainly on rotting vegetation, bacteria, fungi, and animal remains. They rarely eat living plants.*

# WOW!

The creature with the biggest appetite is a moth caterpillar that eats 20 times its own body weight each day.

## FASTEST FEEDERS

Some animals capture their food in less than the blink of an eye. The chameleon lizard flicks out its long stretchy tongue, grabs a grasshopper on its sticky tip, and pulls it into its mouth – all in less than 1/10th of a second. Any slower and the victim might leap to safety. A chameleon's tongue can be as long or even longer than its body, depending on the species.

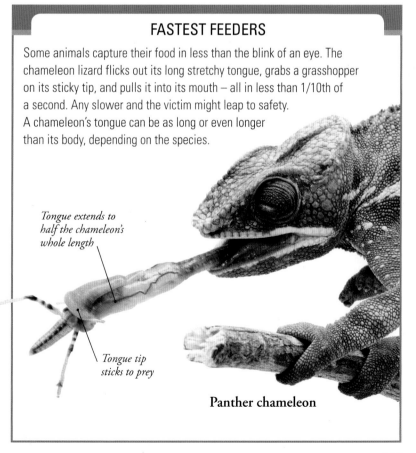

*Tongue extends to half the chameleon's whole length*

*Tongue tip sticks to prey*

**Panther chameleon**

# Predators and prey

In nature's constant battle for survival, predators or hunters must catch their food (prey). Meanwhile, prey try to avoid becoming victims by defending themselves. Usually there is a balance – sometimes predators kill, sometimes prey escape.

## PHYSICAL WEAPONS

Predators use weapons such as sharp teeth to grab, kill, and cut up victims. Some have fangs – long, thin teeth – that may inject venom. Pincers can grab prey to crush or slice. Prey also have weapons, but for self-defence, such as long spines, hard shells, tough scales, and thick skin.

*Spiny quills*

▶ DEFENCE *A porcupine may run at a predator and jab in its spines, called quills, which pull out of the porcupine's skin and stick into the attacker.*

*Large, serrated teeth*

▲ ATTACK *A shark has jagged, saw-edged teeth, jaws that open very wide, and strong muscles to snap them shut with incredible power.*

## SPEED

Predators such as leopards, tigers, and lions are fast runners and use a burst of incredible speed to catch a meal. However, their prey are almost as speedy, and also able to dart and weave suddenly to and fro, to dodge the attack.

▼ ATTACK AND DEFENCE *Animals can use speed to chase their prey as well as to escape from their predators.*

## USING COLOUR

► WARNING COLOURS are bright patterns, such as red and black, yellow and black, and green and black, showing that a creature is poisonous, or bad to eat. This shield bug has bright red and black stripes as warning colours.

► CAMOUFLAGE is looking like the surroundings, to blend in and be less noticeable. It is used by both predators and prey. This crab spider is camouflaged in the same yellow colour as the flower to help it ambush its prey.

## CHEMICAL WEAPONS

Venoms are toxic substances that are jabbed or stabbed into another creature using fangs, stingers, and similar sharp body parts. A venom causes harm, pain, paralysis (being unable to move), and even death.

Poisonous sting

► ATTACK Using its tail stinger, a scorpion may jab venom into prey, which it grabs with its large pincers.

► DEFENCE The bombardier beetle squirts two substances from its rear end that mix to produce a stinging, irritating spray against enemies.

## GROUP SIZE

Compared to one predator alone, a group or pack of predators can bring down more and bigger prey. Similarly, a group of prey, such as a herd or flock, has safety in numbers. There are many eyes, ears, and noses to detect approaching hunters and give the alarm, allowing most of the group to escape.

▲ ATTACK A wolf pack can kill a victim as large as an adult deer, which is many times bigger than one wolf. But the meal must then be shared so there may not be much for each wolf.

▲ DEFENCE Reindeer are very fast and graze in large herds, where they are always alert to danger.

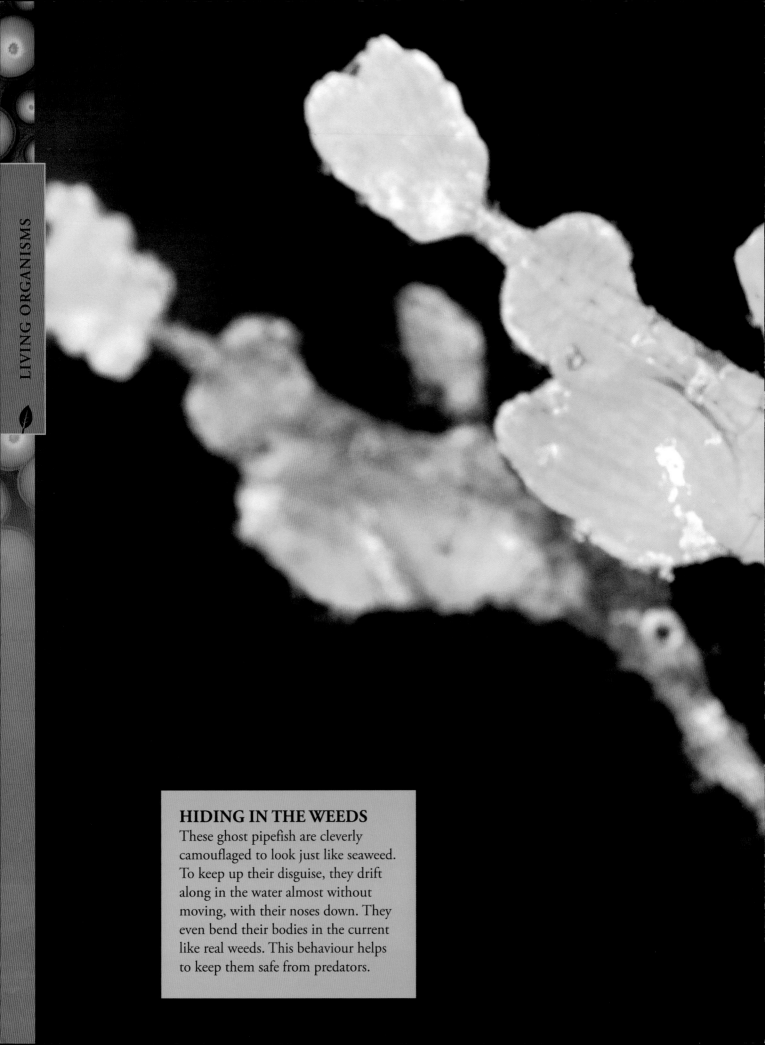

## HIDING IN THE WEEDS

These ghost pipefish are cleverly camouflaged to look just like seaweed. To keep up their disguise, they drift along in the water almost without moving, with their noses down. They even bend their bodies in the current like real weeds. This behaviour helps to keep them safe from predators.

# Senses

The human body's main five senses are sight, hearing, smell, taste, and touch. Many animals have these senses too. Some have fewer, such as underground burrowers or deep-sea dwellers who need no eyes because there is no light there. Other creatures have much more aware senses than ours, and some even have extra senses, such as detecting electricity.

## SIGHT

Our eyes see all the colours of the rainbow, but not all light-type energy. Ultraviolet waves are shorter than ordinary light waves and are seen by various animals including insects, fish, and some birds and mammals, such as reindeer. Some flowers have petal markings that show up in ultraviolet light and guide insects to their sugary nectar.

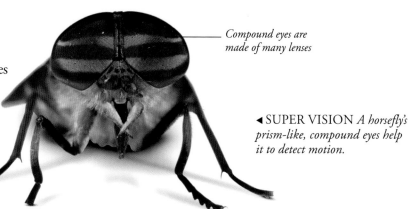

*Compound eyes are made of many lenses*

◀ SUPER VISION *A horsefly's prism-like, compound eyes help it to detect motion.*

## TASTE

Most vertebrates have tongues – muscly organs in their mouths – which they use to manipulate food. Mammals' tongues are covered in tiny sensors called taste buds, which can identify chemicals that are dissolved in saliva. This helps them to recognize what is in their food, and to avoid eating anything rotten or contaminated. Other vertebrates such as fish and reptiles have fewer taste buds, and rely on other senses to identify safe foods.

▶ TONGUE POWER *Tongues are not just for tasting, they also help animals to eat. Lions' rough tongues scrape tiny scraps of meat from the remains of their prey.*

## SMELL

Compared to a human's nose, a dog's long nose has more room for many millions more smell detectors. For certain scents, this makes a dog's sense of smell more than 100,000 times better than our own.

▲ BIG NOSE *Some breeds of dog have more than 200 million microscopic smell detectors inside their noses, compared to about 5 million in a human's.*

## TOUCH

Many animals have sense receptors in their skin so they can feel what is around them. Humans can detect gentle touch, hard pressure, heat and cold, and pain through their skin. Other animals, such as cats, have extra-sensitive whiskers that can pick up the slightest pressure.

▶ STAR-NOSED MOLE *This furry creature tracks underground prey using its sense of touch. The tiny "fingers" on its nose are extremely sensitive.*

## HEARING

Bats fly and find their way even in complete darkness, using sounds. They send out very high-pitched, or ultrasonic, clicks and squeaks. These bounce or echo off nearby objects. The bat hears the returning echoes and works out the object's size, shape, distance, and direction, in a process known as echolocation.

▲ BIG EARS *A bat sends out ultrasonic signals from its nose and mouth, and the huge ears hear the faint echoes.*

WOW!

In clear Arctic air, a polar bear can detect the smell of a dead seal or whale 5 km (3 miles) away.

## EXTRA SENSES

Human beings rely on five main senses to take in information about the world around them. Other animals, however, can recognize information that passes us by, such as tiny changes in the temperature or pressure of the air around them, or detect magnetic or electrical fields in their surroundings.

▲ HEAT SENSE *Pit vipers have bowl-shaped, heat-detecting parts called pit organs, one under each eye. The pits receive heat or infrared rays from a warm object, such as a bird or mammal, and tell the snake its direction, distance, and size.*

▲ ELECTRICAL SENSE *Living bodies make small electrical pulses that spread out through water. The hammerhead shark is especially good at sensing electricity. It does this with the help of tiny pits covering its snout and the underside of its head.*

# Communication

The natural world is busy with all kinds of animals sending messages and signals to each other. They use sight, sound, smell, taste, and touch – often all together. Some messages are simple and understood by many creatures. A hiss or growl says: "If you come near, I'll attack!" Other communications are less obvious and only understood by animals of the same kind.

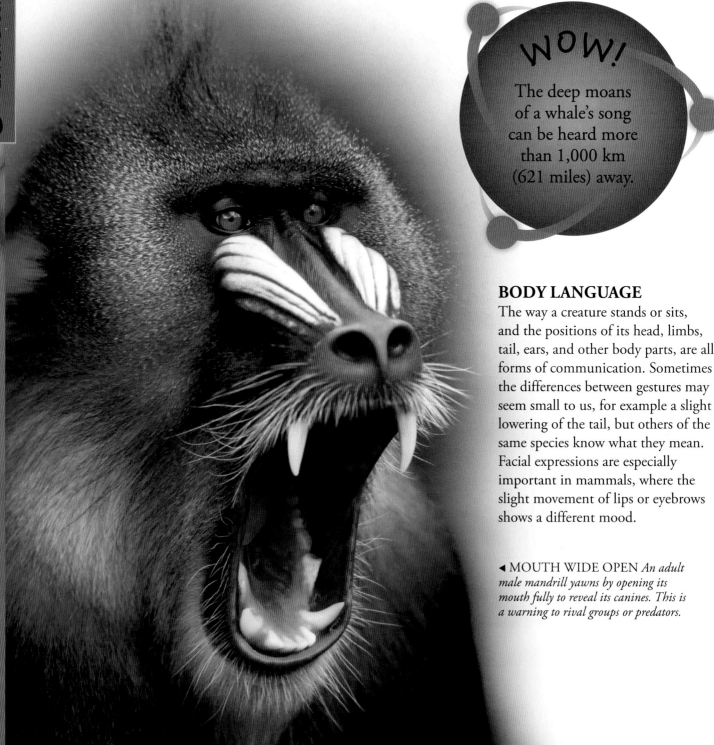

**WOW!**

The deep moans of a whale's song can be heard more than 1,000 km (621 miles) away.

## BODY LANGUAGE

The way a creature stands or sits, and the positions of its head, limbs, tail, ears, and other body parts, are all forms of communication. Sometimes the differences between gestures may seem small to us, for example a slight lowering of the tail, but others of the same species know what they mean. Facial expressions are especially important in mammals, where the slight movement of lips or eyebrows shows a different mood.

◄ MOUTH WIDE OPEN *An adult male mandrill yawns by opening its mouth fully to reveal its canines. This is a warning to rival groups or predators.*

## DEFENDING TERRITORY

Some animals have a territory where they live, feed, and breed. They defend the territory by showing off to and chasing away others. Animals also display and even fight to be in charge of a group such as a herd, and to scare away rivals at breeding time.

▲ GET OUT! *Male impala stamp, snort, and shake heads to scare off invaders in their territory. If this fails, a battle follows.*

## MATING DISPLAY

Most creatures have a way of attracting a mate for breeding. Usually it is the male who puts on the display, for example by dancing and calling. He shows the female that he is fit and healthy, and as a father he will pass on these features to their offspring.

▲ RED BALLOON *The male frigatebird blows up his scarlet throat patch like a balloon to impress the female.*

## WARNING SIGNALS

Defence messages usually involve an animal making itself look bigger and more frightening to an enemy. Mammals make their fur stand on end and birds fluff out their feathers. Often the animal opens its mouth, jumps about, and makes noises to scare the predator.

▲ LOOK BIGGER *The frilled lizard opens out its frill of skin, shows its teeth, and hisses to communicate that it will fight back if attacked.*

### SHARING INFORMATION

Social animals live together with their own kind in groups or colonies. There are many forms of communication among group members, for instance about the location of food, getting ready to move on, or approaching danger. Ants communicate by touch and scents. They give out different scents, or pheromones, for food, attack, defence, and danger.

**A large group of army ants working together.**

251

# Living together

Creatures that live with others of their kind are known as social animals. They sometimes live together in one big home, called a colony. Living together offers many benefits, such as being able to share food and protect each other from predators. Different animals form different kinds of group to give them the best chance of survival.

## TOGETHER FOR LIFE

Several kinds of mammals and birds mate for life. This means choosing and then staying with the same breeding partner for as long as the two survive. Each year the pair renew their close bond and use their combined experience and knowledge to raise their family.

◄ PERMANENT PAIR *Scarlet macaws can stay together for more than 40 years. When one dies, the other may take years to find a new partner.*

## WOW!

Army ants live and move around in huge colonies, but they do not build permanent nests. They cling together in large clusters to rest.

## BATTLE FOR MATES

When breeding time arrives in the animal kingdom, males may hold contests to attract the best or most partners. The males show off their health and readiness to mate. The biggest battles are between male elephants, who head-butt and clash tusks, sometimes causing deep wounds.

▼ SHAKING THE GROUND
*Male elephants spread their ears, trumpet, stamp, and shove to decide which one gets to breed.*

## SMALL GROUPS

Social groups of 10 to 20 are common among smaller animals. In a group of meerkats, also known as a "mob" or "gang", only one pair – the alpha female and male – produce young. Some less important females baby-sit and feed the offspring, while others contribute to the group by watching for danger, communicating food sources, and sounding alarm calls.

▶ WATCHING OUT *Adult meerkats often climb to a higher point to spot potential predators. If a predator is detected they warn the rest of the group.*

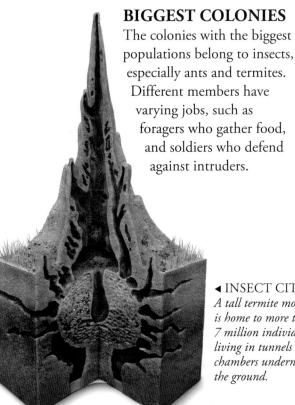

## BIG GROUPS

The biggest animal groups include bird flocks of tens of thousands, fish shoals with hundreds of thousands, and swarms of locusts containing many millions. These vast gatherings may travel to find food, or to breed, or simply stay together to discourage predators.

◀ AIR TRAFFIC *While they feed in smaller groups by day, starlings gather at dusk into groups of thousands, sometimes even millions.*

## BIGGEST COLONIES

The colonies with the biggest populations belong to insects, especially ants and termites. Different members have varying jobs, such as foragers who gather food, and soldiers who defend against intruders.

◀ INSECT CITY
*A tall termite mound is home to more than 7 million individuals, living in tunnels and chambers underneath the ground.*

### BEING SOCIAL

The most complex mammal societies include dolphins, whales, lemurs, monkeys, apes, and elephants. They have many ways of communicating complex information (see p.250) about who is in charge, food sharing, friendships, rivalries, and mating. As young chimps play and rest together, they learn hundreds of signals and messages.

**Young chimps watch adults to learn new behaviour.**

# Migration and hibernation

Many regions of the world have a warm season each year with plenty of food and good living conditions, followed by a cold season when survival is much more difficult. There are two main ways of staying alive through the difficult times – stay or go. That is, stay and enter hibernation or torpor, or leave on migration.

▶ FLYING FAR *The tiny ruby-throated hummingbird flies between North-East and Central America to escape the cold weather. Some of them take an almost non-stop journey of 1,000 km (620 miles).*

## MIGRATION

Migration is a long-distance journey, usually made around the same time each year along a regular route. Most animals migrate away from a cold place to where it is warmer, then back again, each year. A huge variety of creatures migrate, from worms, lobsters, and insects to fish, birds, and mammals.

◀ ON THE HOOF *Caribou, or reindeer, travel to where their food is available. In summer, they trek north to find seasonal plants to eat. In winter they travel back south where it is warmer.*

▶ LONG-DISTANCE SWIMMERS *Atlantic salmon travel from freshwater streams to the Atlantic Ocean for food. After a few years, they return to the fresh waters to lay eggs.*

▲ MONARCH ROOST *Monarch butterflies fly up to 4,800 km (3,000 miles) from their summer feeding grounds to their winter roosts.*

## TOO COLD TO MOVE

Cold-blooded animals, such as reptiles, fish, and insects, do not truly hibernate. Since they cannot make warmth in their bodies, they become so cool in winter that their muscles can no longer work. This inactive condition is called torpor, and is usually for a shorter period than hibernation. Some insects may even freeze solid, then thaw out and continue living when it warms up in spring.

◄ GARTER SNAKES *These snakes go into torpor in large numbers in common dens during winters.*

◄ WOOLLY BEAR CATERPILLAR *These caterpillars survive the extreme cold of the Arctic by letting their bodies freeze.*

## HIBERNATION

Hibernation is a form of inactivity in which animals enter deep sleep to survive the harsh winters. Warm-blooded animals, mainly mammals, hibernate by hiding away and lowering their body temperatures by up to 30°C (50°F). Breathing, heartbeat, and other body processes become very slow to save energy.

▼ IN A NEST *Hibernation is much deeper than normal sleep and the animal, such as this dormouse, cannot wake up easily.*

▲ IN A CAVE *Bats cluster in groups to hibernate in caves, where the temperature is cold but it rarely freezes.*

# Habitats and ecosystems

A habitat is a particular kind of place or surroundings, where something lives. Together, all the living things in each habitat are known as a community. These are plants and animals including herbivores, carnivores, and detritivores (see pp.242–243). An ecosystem describes the ways in which these living things interact, for example, who eats what in the food chain.

## MAJOR HABITATS

Physical conditions such as the hardness of rock or saltiness of water, the land's height or the water's depth, and the yearly climate and seasons, all greatly affect which plants and animals live in a habitat. Warmth, moisture, deep soil, and lack of wind all encourage life. Hard rocks, extreme temperatures, and lack of water make survival more difficult.

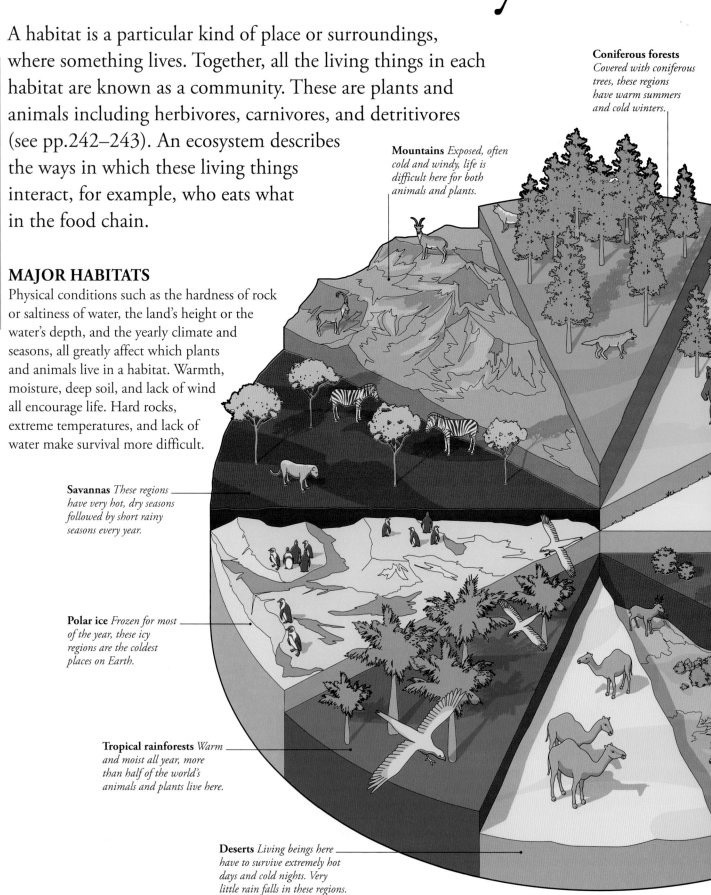

**Coniferous forests**
*Covered with coniferous trees, these regions have warm summers and cold winters.*

**Mountains** *Exposed, often cold and windy, life is difficult here for both animals and plants.*

**Savannas** *These regions have very hot, dry seasons followed by short rainy seasons every year.*

**Polar ice** *Frozen for most of the year, these icy regions are the coldest places on Earth.*

**Tropical rainforests** *Warm and moist all year, more than half of the world's animals and plants live here.*

**Deserts** *Living beings here have to survive extremely hot days and cold nights. Very little rain falls in these regions.*

# FOOD CHAINS AND WEBS

Food chains show examples of who eats what in a particular habitat. Each chain starts with a plant, then a herbivorous animal, a carnivore, and ends with a top carnivore – an animal that few others prey on. In most habitats, animals have varied diets so food chains link together into larger food webs.

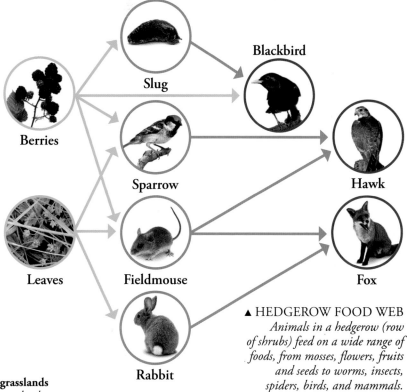

▲ HEDGEROW FOOD WEB *Animals in a hedgerow (row of shrubs) feed on a wide range of foods, from mosses, flowers, fruits and seeds to worms, insects, spiders, birds, and mammals.*

## WOW!

The coldest habitat is the Antarctic ice cap, where emperor penguins cope with temperatures of –50°C (–58°F).

**Deciduous forests** *These regions experience four distinct seasons. The trees grow and lose their leaves every year.*

**Temperate grasslands** *These dry, grassy lands support grazing herds such as bison, asses, or kangaroos.*

**Mediterranean** *Filled with scrub vegetation, these regions have hot, dry summers, and much milder, wetter winters.*

**Tundra** *Cold and treeless, the summer months attract animals such as reindeer and nesting birds.*

# BIODIVERSITY

The more different kinds of living things there are in a habitat, the greater its biodiversity. Tropical forests and coral reefs are the most biodiverse with hundreds, even thousands, of species in each small area. Polar regions, deserts, and the deep-sea bed are much less biodiverse.

▲ MOST VARIED *A coral reef has a huge range of species living in mini-habitats such as rocky caves and burrows.*

# Ecology and conservation

Ecology is the study of how living things fit together in their surroundings. For example, what are the predators of a certain animal? Which herbivores eat a plant's fruits and spread its seeds? What kinds of nest sites do animals need? Sadly, more wild places are destroyed every year. Knowing their ecology helps us to work out how we can save or conserve them for the future.

## MAIN THREATS TO WILDLIFE

- Habitat loss, such as logging, burning, and clearance for human use
- Global warming, especially affecting coral reefs by killing polyps
- Pollution, such as oil slicks, pesticide sprays, and industrial wastes
- Poaching and hunting for food and trophies such as tiger bones and rhino horns
- Diseases, such as chytrid fungus killing amphibians
- Invasive species replacing native ones

**Habitat destruction through deforestation**

**Coral death due to global warming**

**Imported species kill off native wildlife**

## WILD LAND TO FARMLAND

Massive areas of grasslands and other natural habitats are taken over each year for farming – to grow crops and to raise cows, sheep, and other domestic animals. The wild animals who lived there are forced into smaller areas and die from lack of food, disease, and other problems of overcrowding.

▼ NOWHERE TO GO *As people fence off more African grasslands to grow crops, natural grazers such as wildebeest and zebras can no longer follow their age-old migrations (the regular movement of animals from one place to another).*

## THE NEED TO KNOW

Gathering knowledge about different ecosystems helps to identify the best way to save species and habitats. This is especially important for animals that travel widely, such as birds, whales, and turtles. Knowing the places they visit and the risks they face leads to a conservation plan of how best to keep them alive.

◄ KEEPING TRACK *This hawksbill turtle has a small tracking device fitted that records where it swims to feed, rest, mate, and lay eggs.*

## WOW!

One in five mammal species, and one in three amphibians, are on the official lists of threatened animals.

## CAPTIVE BREEDING

Breeding creatures in wildlife parks and zoos can help to save a very rare species from dying out. It also raises people's awareness of the need for conservation. The aim in the long term is to release the animals back into their safe wild habitat.

► SOMEWHERE TO LIVE *Baby giant pandas are born in breeding centres because their natural habitat is endangered.*

## CONSERVATION CAMPAIGNS

Saving big, spectacular wild animals, such as gorillas and dolphins, depends on preserving their habitats – which also saves all the other creatures and plants living there. Governments and many organizations work to protect important habitats.

► SAVED *Mountain gorillas live in only a few small highland areas of Central Africa. With huge conservation effort, their numbers have increased in recent years.*

▲ LOST *The rare Chinese river dolphin, also known as baiji, suffered from pollution, overfishing, lack of food, and other hazards. In 2010 it was declared extinct.*

# The human body

The human body has more than 1,000 parts – including bones, muscles, and other organs – all working together every second of every day. The body is fuelled by food substances, especially starches, sugars, and fats. These combine with oxygen, taken in by the lungs, to release energy that keeps the body warm and powers its movements and processes.

## UNDER THE SKIN

The body's main parts, such as the skin, brain, eyes, heart, and stomach, are called organs. Each contains millions of busy cells. The organs fit closely together and are always working – the heart beats, the lungs breathe, the intestines process food, and the muscles move our body.

▶ BRAIN *Filling the top half of the head, the brain controls movements and is the place for thoughts, emotions, and memories.*

▶ HEART *Beating once every second or more, the muscular heart sends blood out through blood vessels (tubes) that reach every body part.*

*The whitish enamel covering the teeth is the hardest substance in the body.*

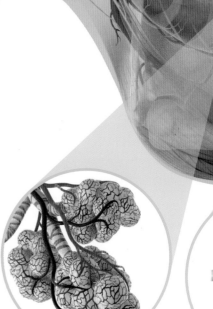

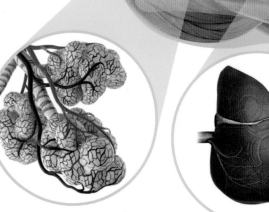

▶ LUNGS *Deep in the lungs are millions of bubble-like alveoli, which take in oxygen from air and get rid of carbon dioxide.*

▶ LIVER *The largest inner organ, the liver takes substances from digested food and alters them. It stores some and releases others into the blood.*

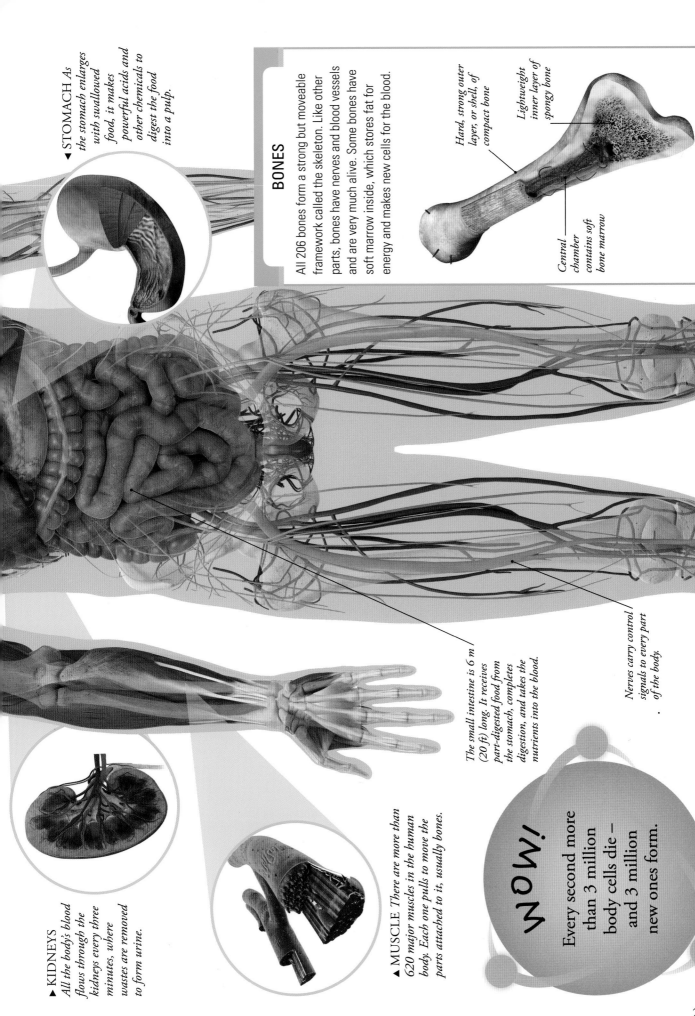

▼ STOMACH *As the stomach enlarges with swallowed food, it makes powerful acids and other chemicals to digest the food into a pulp.*

## BONES

All 206 bones form a strong but moveable framework called the skeleton. Like other parts, bones have nerves and blood vessels and are very much alive. Some bones have soft marrow inside, which stores fat for energy and makes new cells for the blood.

*Hard, strong outer layer, or shell, of compact bone*

*Lightweight inner layer of spongy bone*

*Central chamber contains soft bone marrow*

*The small intestine is 6 m (20 ft) long. It receives part-digested food from the stomach, completes digestion, and takes the nutrients into the blood.*

*Nerves carry control signals to every part of the body.*

▲ KIDNEYS
*All the body's blood flows through the kidneys every three minutes, where wastes are removed to form urine.*

▲ MUSCLE *There are more than 620 major muscles in the human body. Each one pulls to move the parts attached to it, usually bones.*

WOW!
Every second more than 3 million body cells die – and 3 million new ones form.

261

# Body systems

The body's main organs and tissues work together in groups called systems. Each individual system carries out an important task to keep the whole body alive and healthy. For example, the bones and joints form the skeletal system for support, and the muscular system keeps the heart beating and the body moving.

## DIGESTIVE SYSTEM

This system starts in the mouth, as teeth chop up and chew food, then mix it with saliva (spit) so that it is moist and easily swallowed. At the other end, leftovers of digested food come out of the opening at the anus.

## CIRCULATORY SYSTEM

The three main parts of this system are the muscular pump of the heart, the network of blood vessels, and the blood that passes through them. Blood delivers oxygen, energy, and nutrients, and collects wastes for disposal.

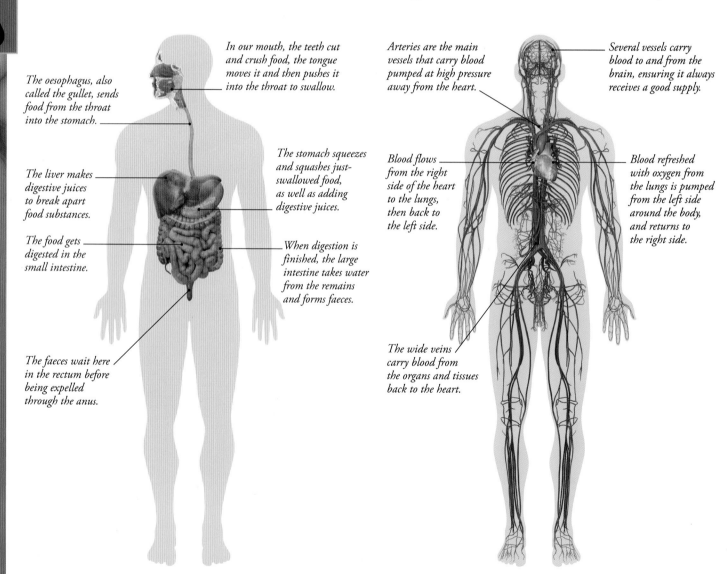

*The oesophagus, also called the gullet, sends food from the throat into the stomach.*

*In our mouth, the teeth cut and crush food, the tongue moves it and then pushes it into the throat to swallow.*

*The liver makes digestive juices to break apart food substances.*

*The stomach squeezes and squashes just-swallowed food, as well as adding digestive juices.*

*The food gets digested in the small intestine.*

*When digestion is finished, the large intestine takes water from the remains and forms faeces.*

*The faeces wait here in the rectum before being expelled through the anus.*

*Arteries are the main vessels that carry blood pumped at high pressure away from the heart.*

*Several vessels carry blood to and from the brain, ensuring it always receives a good supply.*

*Blood flows from the right side of the heart to the lungs, then back to the left side.*

*Blood refreshed with oxygen from the lungs is pumped from the left side around the body, and returns to the right side.*

*The wide veins carry blood from the organs and tissues back to the heart.*

▲ WHERE DIGESTION HAPPENS *The main organs of the digestive system form one long passageway, the gut. They fill most of the lower half, or abdomen, of the main body.*

▲ BODY-WIDE NETWORK *Blood vessels branch out into every body part. Added together end-to-end they would stretch more than 100,000 km (62,000 miles).*

## SKIN REPAIR

When your skin is cut, blood is always ready to mend it. Special proteins in blood form a tangle of threads, or fibres, that trap blood cells. At the same time, tiny cells in the blood, called platelets, come together and become sticky. These two processes help form a clot, that hardens and stops bleeding.

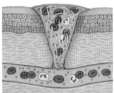 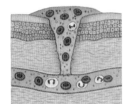 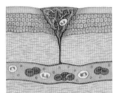 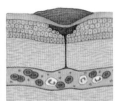

*1. Blood leaks from a fresh cut*

*2. Fibres in blood trap blood cells to form clot*

*3. Clot stops leak and begins to harden*

*4. Clot solidifies into scab and protects new skin beneath*

## NERVOUS SYSTEM

The brain is linked by nerves to every corner of the body. Some nerves carry messages to the brain, from the eyes, ears, and other senses. Other nerves take messages from the brain to the muscles, to control movements.

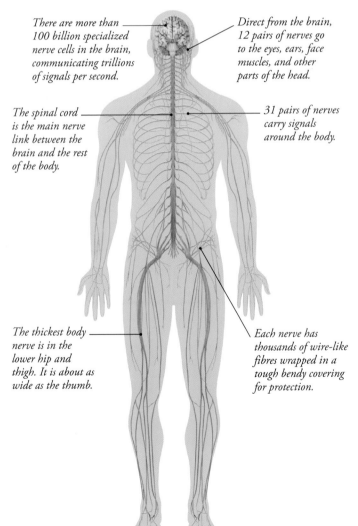

*There are more than 100 billion specialized nerve cells in the brain, communicating trillions of signals per second.*

*Direct from the brain, 12 pairs of nerves go to the eyes, ears, face muscles, and other parts of the head.*

*The spinal cord is the main nerve link between the brain and the rest of the body.*

*31 pairs of nerves carry signals around the body.*

*The thickest body nerve is in the lower hip and thigh. It is about as wide as the thumb.*

*Each nerve has thousands of wire-like fibres wrapped in a tough bendy covering for protection.*

▲ INFORMATION WEB *Nerves send and receive messages in the form of tiny pulses of electricity, thousands every second, some travelling more than 100 metres per second (328 feet per second).*

## REPRODUCTIVE SYSTEM

The female reproductive system makes tiny egg cells, and the male system produces even smaller sperm cells. When a sperm and egg join, the resulting fertilized egg develops in the womb inside the female body. Nine months later a baby is ready to be born.

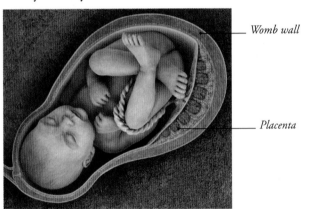

*Womb wall*

*Placenta*

▲ IN THE WOMB *The placenta helps to transfer oxygen and nutrients from the mother's blood to the baby's.*

## BODY DEFENCES

The body contains more than 50 billion white cells. Most are in the blood but they also squeeze out into the gaps between other cells and tissues. The white cells defend the body by attacking any invaders and other dangers, from microscopic germs to parasitic worms.

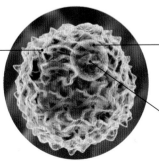

*White blood cells engulf and "eat" bacteria*

*Folds of white cell surround and enclose bacteria*

*Bacteria are taken in, digested and destroyed*

**White blood cell destroying bacteria**

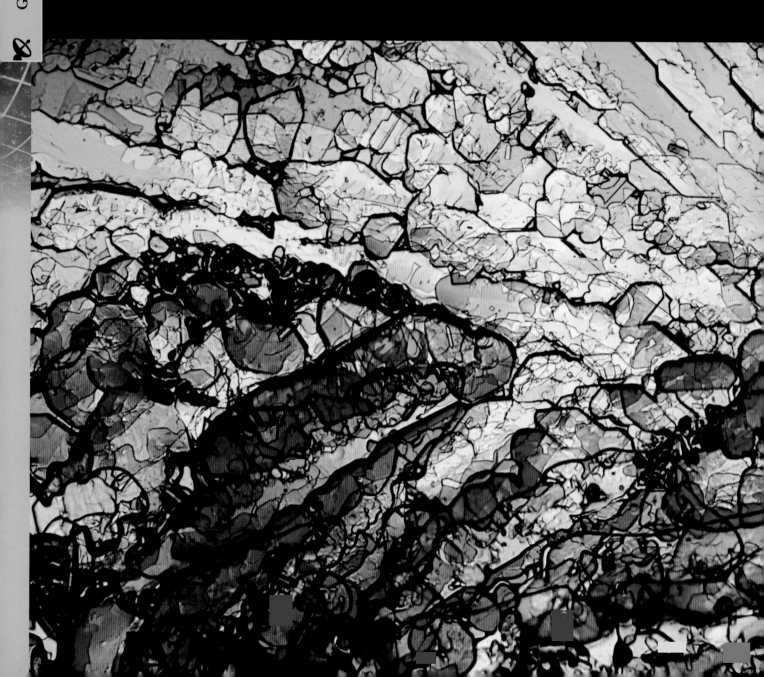

# GREAT
# DISCOVERIES

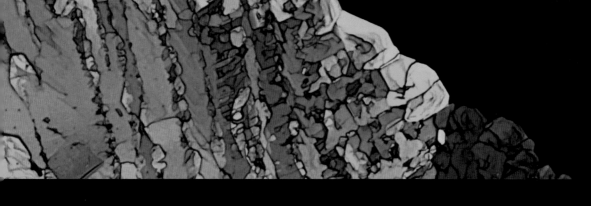

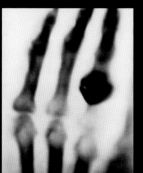

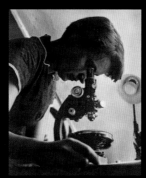

For thousands of years, scientists have tried to understand how the world works. Their amazing discoveries – some by accident – have transformed the way we live.

**LIFE-SAVING CRYSTALS**
These colourful swirls are crystals of a substance called insulin. In the 1920s, scientists discovered how to use insulin to treat a dangerous disease called diabetes. Its discovery has saved countless lives.

# Progress in science

The earliest humans knew very little about the world. But they did know that some berries were tasty while some were poisonous, and that some rocks could be used to make fire or tools. As civilization advanced, people learned to use these observations to come up with new ideas. Later on, they learned to test these ideas, and use the results to bring about new technologies. Science is a continuous learning process – the more we learn, the more ways we can make our lives better.

The rover's cameras can send detailed images back to Earth

Solar panels use sunlight to power the rover

## SCIENCE IN ACTION

Advances in science have helped us to achieve amazing things. We can cure diseases that have killed millions of people, send videos and messages around the world in seconds, and build machines that can detect particles smaller than atoms. We have even sent machines to explore other planets in our Solar System. But there is still plenty for us to discover, on our own planet and beyond.

◄ MARS ROVER *This rover is one of two probes that landed on Mars in 2004 to examine the planet's rocks, soil, and atmosphere.*

## A FAIR TEST

The most important tool in science is the experiment. Scientists use these special tests to see whether their ideas are correct. For an accurate test, every possible influence must be controlled, so that we can be sure any changes are due to the test and not other factors. The experiment below tests how quickly different types of beans grow. Only the type of bean was changed. The plants were all given the same time to grow, and the same amount of light and food, to ensure the test was fair.

*This type of bean grows faster than the others in the same conditions.*

*All the beans had the same material to grow on.*

## THE SCIENTIFIC METHOD

Scientists work by observing events in the world around them (known as phenomena). They then ask a question – why or how does this happen? Then they come up with a hypothesis, or an idea, that answers that question. Finally, they conduct an experiment that tests their hypothesis. If the experiment is a success, the hypothesis is taken as true. If not, the hypothesis is rejected and the scientist looks for a new explanation.

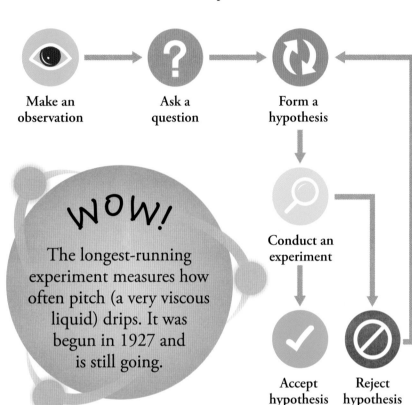

**Make an observation** → **Ask a question** → **Form a hypothesis** → **Conduct an experiment** → **Accept hypothesis** / **Reject hypothesis**

**WOW!**

The longest-running experiment measures how often pitch (a very viscous liquid) drips. It was begun in 1927 and is still going.

## ACCIDENTAL DISCOVERIES

Sometimes accidents or luck lead people to new ideas and new techniques. One such example is the discovery of antibiotics – drugs used to kill bacteria that cause diseases. The first antibiotic, penicillin, was accidently discovered when a scientist allowed mould to grow in a dish of bacteria he was growing for other experiments. He noticed that the area around the mould was free of bacteria – the mould was producing a substance which killed off the germs.

◄ VULCANIZATION *Charles Goodyear invented a new technique for hardening rubber when he accidentally spilt some samples on to a hot stove.*

▲ PENICILLIN *This powerful germ-killing bug was accidently discovered by the Scottish scientist Alexander Fleming in 1928.*

# Ancient science

We take for granted many tools we use today, but all of them – even the very simplest – had to be invented somewhere. People from ancient civilizations across the world discovered and wrote down some of the earliest scientific ideas, passing them down to us today. Their inventions gave later scientists the basis for new ways of thinking. They also laid down rules for how to think about science and maths that are still used to this day.

## EUREKA!

The Ancient Greeks began to come up with scientific processes in the 5th century BCE. One early scientist, Archimedes, was challenged to find out whether a gold crown was all gold or just lead covered in gold leaf. According to legend, he discovered the solution when he sat down in his bath, slopping water over the edge. This gave him the idea of immersing the crown in water to calculate its volume. He leapt from his bath shouting "Eureka", meaning "I have found it (the answer)".

▲ BATHTIME BRAINWAVE *This medieval drawing shows Archimedes in his bath with gold crowns and iron weights nearby.*

## WATCHING THE SKIES

Ancient civilizations were often interested in the Sun, Moon, and stars. They noticed that the Sun was highest in the sky at the same time every day – noon – but sunrise and sunset were at different times each day. This gave rise to the idea of seasons. They noticed that the Moon changed shape in a regular pattern lasting 28 days, giving us the idea of months. And they tracked the motion of stars, watching how they moved through the sky at different times of year, and even picking out the other planets in our Solar System.

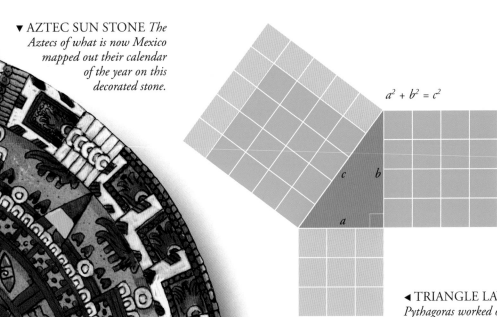

▼ AZTEC SUN STONE *The Aztecs of what is now Mexico mapped out their calendar of the year on this decorated stone.*

$$a^2 + b^2 = c^2$$

*c*  *b*

*a*

## NUMBERS AND MATHS

As well as observing nature, early scientists began to work with shapes and numbers – the beginnings of mathematics. The works of the Greek philosopher Euclid, written around 300 BCE, were used to teach maths across Europe for more than 1,000 years. Another Greek, Eratosthenes, calculated the distance around the Earth almost perfectly, 1,700 years before modern scientists.

◄ TRIANGLE LAW *The Greek philosopher Pythagoras worked out that, for any triangle with a right angle, the square of side "a" and the square of side "b" would equal the square of side "c".*

## BUILDING SHAPES

Ancient civilizations used their knowledge of materials and geometry to create astonishing buildings. The ancient Egyptians and Babylonians used right angles to build pyramids, while the Romans discovered how to build self-supporting arches. Many of these buildings were constructed so strongly that they are still standing today.

▲ ROMAN ARCHES *This amazing structure, called Pont du Garde, in France, was built by the Romans to carry water.*

*Dragon facing the direction of the earthquake drops ball into toad's mouth*

*Eight dragon heads point in different directions*

## SCIENTIFIC INSTRUMENTS

With new ideas about science came new machines and instruments. Some of these were built to help with difficult tasks, such as cranes, ramps, and levers for building works. Others were built to help with scientific studies – from the earliest clocks, which used the position of the Sun to measure time, to devices for working out where an earthquake was taking place.

◄ EARTHQUAKE DETECTOR *In about 100 CE, Chinese scientist Zhang Heng created the first seismograph – an instrument to locate the direction of earth tremors.*

# Islamic science

By the 7th century CE, the ancient civilizations had started to fall apart. New empires sprang up around the world and began to explore their own scientific ideas, often inspired by the discoveries of older cultures. One of the largest empires was formed as Islamic peoples spread across Asia. They valued scientific study and, between the 8th and 12th centuries CE, made great advances in astronomy, mathematics, and medicine.

▲ THE BOOK OF OPTICS *Ibn al-Haytham's ideas about light spread across Europe and Asia. This medieval edition of* The Book of Optics *shows how ibn al-Haytham's ideas were used to explain refraction and reflection.*

## LIGHT IDEAS

Many of the great advances in science came as Islamic scholars improved on the ideas of Ancient Greek philosophers and scientists. Two Greeks, Aristotle and Euclid, had different ideas about how our eyes see. Aristotle believed that light shone out of objects and into our eyes, while Euclid believed that our eyes beamed out rays to help us see. The Islamic scholar ibn al-Haytham used observations and experiments to prove that Aristotle was right.

## THE SCIENCE OF SHAPES

Islamic cultures at this time often decorated their buildings with colourful ceramic tiles. Mathematicians worked on ancient ideas of geometry (the mathematics of shapes and angles) to plan out complex patterns of shapes which fit together perfectly so there were no gaps between the tiles. The mathematics of interlocking (perfectly fitting) shapes is called tesselation.

▶ TESSELATING TILES *These tiles in Morocco fit together perfectly in a neat mathematical pattern.*

## PREDICTING THE STARS

Islamic scholars made great advances in astronomy – the study of the stars. Arab astronomer al-Battānī (858–929) worked out the length of the solar year, which is how long it takes for the Earth to travel around the Sun. He also helped to generate tables of numbers to predict where the Sun, Moon, and planets could be seen in the sky at different times of year. Other scholars, like al-Zarqālī (1029–1087), built instruments called astrolabes to help map out the positions of stars in the sky.

◄ ASTROLABE *These instruments map the movements of the Sun, Moon, planets, and stars.*

## MATHS MASTER

Great developments in mathematics were made during the Islamic Age. The one we use every day is numbers 0–9. A scholar named al-Khwārizmī (780–850) adapted a system of numbers used in India to create Arabic numerals (shown below). It spread across Asia and Europe to form the numbers we use to this day. He also developed new ways of using algebra – a way of applying equations to find out mathematical answers.

| 0 | 1 | 2 | 3 | 4 | 5 | 6 | 7 | 8 | 9 |
|---|---|---|---|---|---|---|---|---|---|
| ٠ | ١ | ٢ | ٣ | ٤ | ٥ | ٦ | ٧ | ٨ | ٩ |

## IBN AL-HAYTHAM

Ibn al-Haytham (965–1040), also known as Alhazen, is often called the world's first scientist. He made careful observations to test his ideas, making sure that all the other conditions of his experiment remained the same so that no other influences could disrupt his results.

## CURING THE SICK

Ancient medicine was often closer to magic than science. Islamic scientists such as ibn Sīnā (980–1037) tried to make the study of medicine more scientific. They made careful lists of the symptoms of illnesses, and experimented with different cures to find out which ones worked and which were based on superstition. They also tried out different mixtures of herbs in different amounts to create medicines that were effective and safe to use.

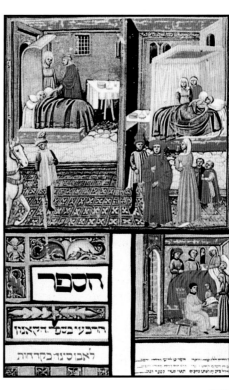

▶ MEDICAL MANUAL The Canon of Medicine *by ibn Sīnā was a huge encyclopedia of medicine, which was used across Europe and Asia for hundreds of years.*

# Planets and pendulums

The 16th century saw a scientific revolution in Europe. A wave of new ideas spread across the continent, taking the discoveries of Ancient Greek, Roman, and Islamic scholars and using them to create new fields of exploration and research. These discoveries changed the way people looked at the world, providing a basis for inventions such as machines and electricity, which make modern life possible.

## ASTRONOMY AND SPACE

One of the most important scientists of the new age was an Italian named Galileo Galilei. A student of mathematics, Galileo went on to develop his own ideas, and was able to build telescopes that could see further and more clearly than before. His writings made him unpopular with powerful members of the Catholic Church, who felt that his new ideas challenged their authority, and he was arrested and imprisoned.

**WOW!**

Galileo was able to estimate the heights of mountains on the Moon by measuring the lengths of their shadows.

▼ SEEING THE STARS *Using his improved telescope, Galileo was the first person to record that the planet Venus changes shape in the sky, just like the Moon.*

## SOLAR SYSTEM AND PLANET ORBITS

The Ancient Greeks believed that the Earth was at the centre of the Universe, and the Sun and all the stars orbited around it. A Polish astronomer named Nicolaus Copernicus, who studied in Italy in the early 16th century, challenged this view. His measurements of the movements of the Sun and planets led him to believe that the Earth orbited the Sun. He also correctly positioned all of the known planets at the time and explained why the seasons occurred. His theories were backed up by Galileo's studies in astronomy around 30 years later.

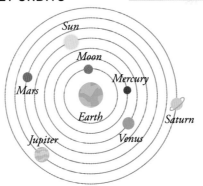

▲ EARTH AT CENTRE *In earlier times it was believed that the Earth sat at the centre of the Universe, and the Sun and planets orbited around it.*

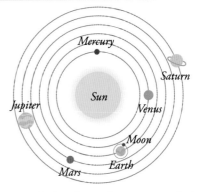

▲ SUN AT CENTRE *Copernicus worked out that the Earth and the other planets orbit the Sun. He also said that the Earth spins on its axis, and the Moon orbits around the Earth.*

## THE PENDULUM'S SWING

Galileo also studied the mechanics of movement, especially the way things fall or swing on a rope. He watched a pendulum swinging and noticed that it always crossed its middle point in time with his heartbeat, no matter how far it swung. Galileo concluded that a pendulum always takes the same time to complete a cycle – swinging from one side to the other. Later scientists used this principle to build the first pendulum clocks.

▲ TOWER TESTS *Legend has it that Galileo began his experiments on pendulums by swinging the chandeliers in the famous Cathedral of Pisa, Italy.*

## THE BEST AIM

Another area in which Galileo made important discoveries was the mathematics of trajectories, or how objects fly through the air when gravity pulls on them. A ball fired from a cannon flies in a steady curve called a parabola, moving up with the force of the cannon shot, then curving back down to Earth as gravity acts on it. Galileo worked out how changing the angle of the cannon changes the distance the cannonball flies when it is fired.

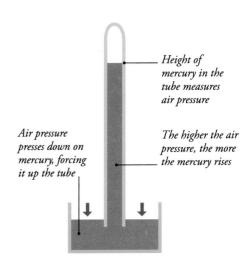

Height of mercury in the tube measures air pressure

Air pressure presses down on mercury, forcing it up the tube

The higher the air pressure, the more the mercury rises

## AIR PRESSURE

Galileo's writings influenced many scientists. Evangelista Torricelli, born in Italy in 1608, read many of Galileo's works and used them to develop his own ideas. He is most famous today for inventing the barometer (above), a machine that measures air pressure using a column of mercury. This device can be used to predict changes in the weather.

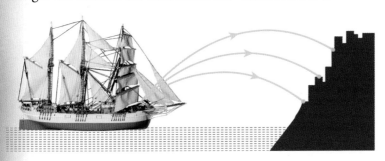

# Gravity and rainbows

In the late 1600s, another wave of scientific discoveries spread across the world. This was known as The Enlightenment, and combined scientific thinking with new ideas, philosophy, and politics. One of the most famous scientists of this period was Englishman Isaac Newton, whose ideas about light, gravity, and motion are still used by scientists and engineers to this day.

## DISTANT ATTRACTION

Newton was fascinated by the movements of the planets. A German astronomer named Johannes Kepler had already worked out that the planets orbit the Sun in oval shapes, called ellipses. Newton realized that these orbits were controlled by an invisible force – gravity – which pulled the planets towards the Sun. Newton also worked out that gravity must pull between all objects, but that larger objects, such as the Sun, pull with more force than smaller objects. Finally, he showed that the force of gravity gets weaker as objects get further apart.

◄ MOON'S ORBIT *Newton's laws of gravity explain how the Moon orbits around the Earth, and the Earth in turn orbits around the Sun.*

## WOW!

Newton said he came up with his theory of gravity after watching an apple fall from a tree.

## COLOURS OF LIGHT

Newton was also very interested in how light travels between objects. He did experiments into refraction (or the way light changes direction when it passes through a solid object), such as a glass prism. He noticed that coloured light stayed the same colour when it passed through a prism, while white light was split into rainbow patterns. This led him to the idea that different colours are created by different kinds of light (what we now know as different wavelengths). He used his understanding of the movement of light to suggest improvements to telescopes and microscopes.

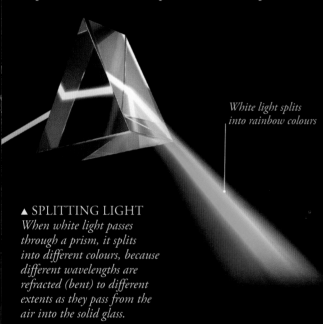

*White light splits into rainbow colours*

▲ SPLITTING LIGHT
*When white light passes through a prism, it splits into different colours, because different wavelengths are refracted (bent) to different extents as they pass from the air into the solid glass.*

▶ CALCULATING CURVES *Calculus lets us work out the area under a curve, and how steep the curve is in different places. This graph shows how fast a motorcycle is going at different times. We can use calculus to find out how far it has travelled, and how quickly it speeds up and slows down.*

## ACTION, REACTION

Newton's three Laws of Motion (see pp. 88–89) are the basis for much of modern mechanics and engineering. The first law says that a stationary object will stay stationary, and a moving one will stay moving, unless a force acts on it. The second law states that the more force acts on an object, the faster it moves. The third law says that any force applied to an object creates an equal force in the opposite direction.

◀ LAWS OF MOTION *Newton's laws explain how a cricket ball moves in one direction until it hits the bat. The ball changes direction when force from the bat is applied to it, and the bat bounces back when the ball hits it.*

## A NEW KIND OF MATHS

Before the 1680s, nobody had discovered how to work out the area under a curved line. Two mathematicians, Newton and a German, Gottfried Wilhelm Leibniz, came up with a new type of mathematics called calculus, which made all kinds of new calculations possible.

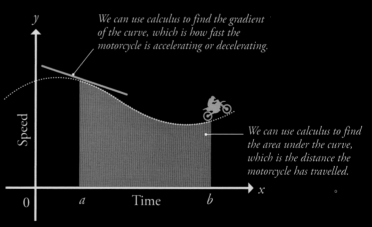

*We can use calculus to find the gradient of the curve, which is how fast the motorcycle is accelerating or decelerating.*

*We can use calculus to find the area under the curve, which is the distance the motorcycle has travelled.*

$y$

Speed

$0$    $a$    Time    $b$    $x$

## RARE VISITORS

Comets are balls of ice and rock which orbit around the Sun. Their orbits take them far out into space, so they are rarely seen from Earth. Thanks to Newton and other astronomers, it became possible to work out the shape of a comet's orbit, and so predict when it would next appear. In 1705, British astronomer Edmond Halley worked out that a certain comet should appear every 75 years, with the next sighting due in 1758. He died before the day arrived, but the comet appeared as he predicted, and has been known as Halley's Comet ever since.

**Halley's Comet**

# Evolution and adaptation

In the 19th century, English naturalist Charles Darwin came up with one of the most important scientific ideas in history. He realized that living things could change across generations, and only the most successful passed their genes on to their children. His theory paved the way for our idea of evolution (see p.222), and showed that all animals and plants, even humans, evolved from the simplest life forms.

### CHARLES DARWIN

Charles Darwin was born in England in 1809. As a young man he began to train as a doctor, but soon became more interested in animals and plants than the human body. His studies in natural history took him all around the world, and led him to come up with radical scientific ideas.

## VOYAGE OF DISCOVERY

In 1831, Darwin joined the crew of HMS *Beagle* on a voyage of exploration and discovery. Over the course of five years, they travelled from England to South America and Australia, crossing the Atlantic, Pacific, and Indian Oceans. The things he saw on this journey provided Darwin with the inspiration for some of his most important ideas.

◄ 1831 DARWIN SETS SAIL *The* Beagle *was sent to map the coasts of the lands of the Southern Hemisphere. Darwin went along as a naturalist, to study local wildlife and bring back samples.*

| 1830 | 1831 | 1832 | 1835 | 1840 |

◄ 1832 MAMMOTH MAMMAL *On arriving in Brazil, Darwin was shown the fossilized remains of a large mammal. He identified it as* Megatherium, *a creature that died out many thousands of years ago, and noted how similar it was to the sloths still living in South America.*

► 1835 ISLANDS OF IDEAS *The* Beagle *spent some weeks exploring the isolated Galapagos Islands off the west coast of Ecuador. Darwin noted that very similar species of birds and tortoises had slightly different body shapes on the different islands. This supported his idea that species change over generations to cope with different conditions.*

## DARWIN'S FINCHES

During his time in the Galapagos Islands, Darwin observed groups of birds with the same appearance but very different beaks. He came to the conclusion that since these islands were far from the mainland and hard for birds to get to, one species of finch had reached them in the distant past. It had evolved into several distinct species, each with a beak specially designed for eating the food on its particular island.

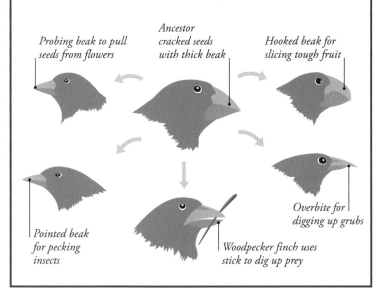

*Probing beak to pull seeds from flowers*

*Ancestor cracked seeds with thick beak*

*Hooked beak for slicing tough fruit*

*Pointed beak for pecking insects*

*Woodpecker finch uses stick to dig up prey*

*Overbite for digging up grubs*

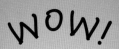

**WOW!**

More than 120 species of animals and plants have been named after Charles Darwin.

◀ 1848 ALLIES AT HOME *After returning to England, Darwin published some of his findings, but continued working on his grand theory in secret. In 1848, another English naturalist, Alfred Russel Wallace, began working on a theory very similar to Darwin's. Together, the two men presented their ideas to other naturalists in 1858.*

▲ 1859 WORD GETS OUT *Darwin's grand theory was finally published in a book called* On the Origin of Species. *It immediately caused great uproar, as many people were unwilling to believe that humans shared ancestors with apes. Cartoons such as this one poked fun at Darwin and his theory of evolution. Despite these attacks, Darwin's idea quickly spread, changing the way we look at life on Earth forever.*

 1848　　1850　　　　　　　　　　　1855　　　1859　1860

## DARWIN'S SUPPORTERS

Many scientists at the time agreed with Darwin's ideas, and used them to help with their own research. Thomas Huxley was one of them. He studied fossils to look for links between prehistoric forms of life and species living today. One of his most famous discoveries was that an extinct winged animal called *Archaeopteryx* had a skeleton very similar to small dinosaurs, but was feathered like a modern bird. He took this as proof that birds had evolved from dinosaurs.

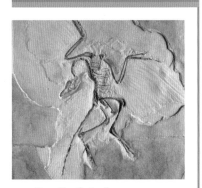

**Fossil of *Archaeopteryx***

# Invisible rays

Towards the end of the 19th century, scientists began to develop delicate instruments for detecting energy signals. They also began to use electricity in their experiments. The result was a series of exciting discoveries, from X-rays (invisible light that can pass right through solid objects) to radioactivity (tiny particles that shoot out of atoms). Together they opened up a whole new world of scientific ideas, full of invisible types of energy that nobody had ever dreamed existed.

## MARIE CURIE

The discovery of radioactivity opened up important areas of scientific research. Some of the earliest and most famous discoveries were made by Marie Curie, a Polish-born scientist working in France. She and her husband, Pierre Curie, discovered two new radioactive elements, polonium and radium, in 1898. After Pierre's death in 1906, Marie worked out new methods for extracting pure radium from its ore, and was the first person to study the metal in detail.

▶ MARIE CURIE AT WORK *This picture shows Curie in her laboratory at the University of Paris, France, where she was the first woman professor.*

### HENRI BECQUEREL

French scientist Henri Becquerel discovered radioactivity by accident in 1896, when he left samples of radioactive uranium next to photographic plates. He noticed that the plates showed dark patches where the uranium had been.

## DISCOVERING X-RAYS

X-rays were discovered in 1895 by a German physicist named William Röntgen. He was investigating the light produced by passing electricity through a vacuum tube. He noticed that, even with a cardboard screen covering the tube, some radiation could still be detected. He named this radiation X-rays, after the scientific term for something unknown, "X".

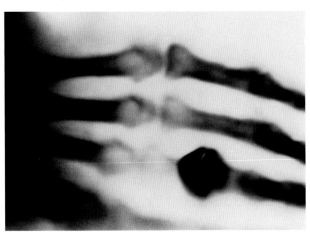

◄ THE FIRST X-RAY *Röntgen took the first X-ray photograph, showing the bones of his wife's hand and her wedding ring.*

## MEDICAL BREAKTHROUGH

Röntgen realized that X-rays could pass through light materials such as paper and skin, but not through denser substances such as lead and bone. This made them very useful for seeing things hidden inside the body, for example broken bones. Today we use X-rays for lots of different purposes, from checking for diseases inside the body, to studying radiation from distant stars.

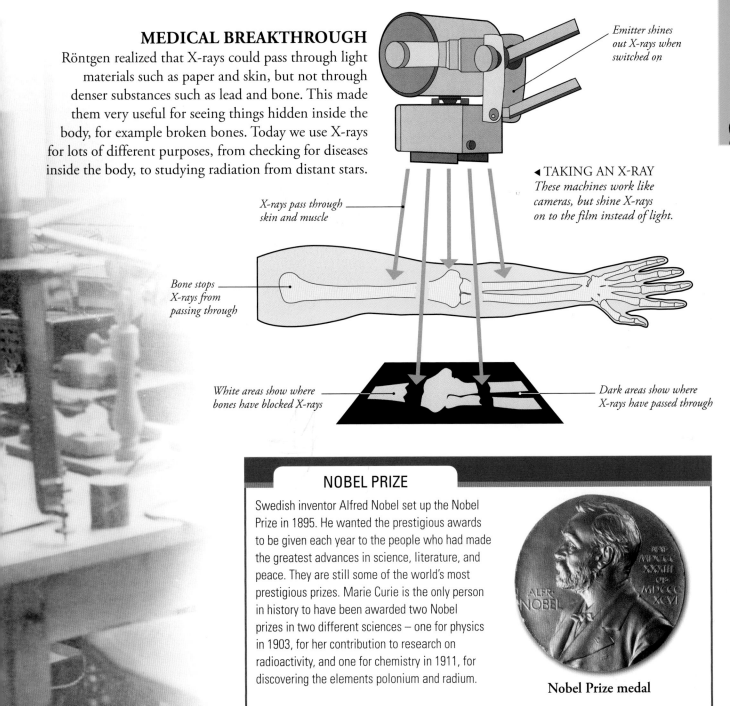

Emitter shines out X-rays when switched on

◄ TAKING AN X-RAY
*These machines work like cameras, but shine X-rays on to the film instead of light.*

X-rays pass through skin and muscle

Bone stops X-rays from passing through

White areas show where bones have blocked X-rays

Dark areas show where X-rays have passed through

### NOBEL PRIZE

Swedish inventor Alfred Nobel set up the Nobel Prize in 1895. He wanted the prestigious awards to be given each year to the people who had made the greatest advances in science, literature, and peace. They are still some of the world's most prestigious prizes. Marie Curie is the only person in history to have been awarded two Nobel prizes in two different sciences – one for physics in 1903, for her contribution to research on radioactivity, and one for chemistry in 1911, for discovering the elements polonium and radium.

**Nobel Prize medal**

# It's all relative

One of the most famous scientists of all time, German physicist Albert Einstein, transformed the way we think about the Universe. At the beginning of the 20th century, his new ideas about maths and physics made scientists across the world think differently about time, energy, and matter. His ideas were so far ahead of his time that we are still working on them today, and he became so famous that his name has been used to mean "genius" ever since.

### ALBERT EINSTEIN

Einstein grew up in Germany and Switzerland. His first job was as a clerk in the Swiss patent office, where he read many scientific papers on topics such as electromagnetism. He began to write his own scientific papers in his spare time. In 1905, at the age of just 26, he published four articles that were read by scientists around the world. The papers investigated the structure of atoms, the movement of electrons, and the relationship between matter and energy. He soon became famous, and won the 1921 Nobel Prize for physics.

◄ GENIUS AT WORK *Einstein moved to the USA in 1933, and worked at the Princeton University for more than 20 years.*

### BROWNIAN MOTION

In 1827, a Scottish botanist named Robert Brown noticed that grains of pollen floating in water seemed to jump about when viewed under a microscope. In 1905, Albert Einstein provided an explanation. Water is a liquid, made up of water molecules that can move freely. The pollen grains are jostled around by the water molecules, which bump into them and push them around. Einstein used this as proof of the existence of atoms and molecules.

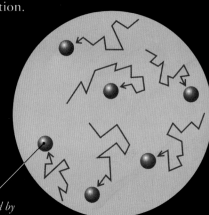

*Grains of pollen dance around as they are pushed by free-moving water molecules*

NIELS BOHR

Danish physicist Niels Bohr worked on some of the same problems as Einstein at around the same time. Bohr worked out that electrons orbit atoms at set distances called shells, and that they can move from a closer orbit to one further away if given energy.

## SPACE AND TIME

A fast object seems to move slowly if you are moving fast as well. It can even seem to move backwards if you overtake it. Einstein showed that this is not true of light, which always moves away from you at light-speed, however fast you travel. This creates an effect known as "time dilation", which means that time slows down for things that travel very quickly. Einstein worked out that time is in fact just another dimension, like the three dimensions of space.

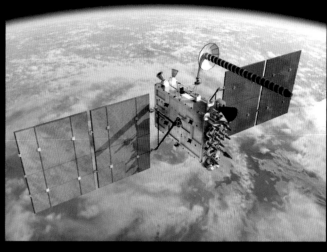

▲ RELATIVITY *GPS satellites rely on very accurate clocks to help people on the ground find out where they are. Because the satellites are moving very quickly in orbit, time runs slightly more slowly for them than for clocks on Earth, so computers are used to correct the signals and keep the system accurate.*

*Sun distorts space, creating a "gravity well"*

*Planet trapped by "slope" in space*

*Space is shown as a flat sheet in this imaginary view*

## MATTER AND ENERGY

As things travel faster, they gain more energy. Einstein calculated that as an object gains energy it also gets heavier, although it takes a huge amount of energy to cause a small increase in mass. His famous equation for this is $E = mc^2$ – that is energy is equal to mass times the speed of light squared (a very big number). From this, Einstein was able to prove that matter can be converted to energy, and vice versa. In fact, matter and energy can be seen as two sides of the same coin.

▲ BENDING SPACE
*According to Einstein, large bodies of mass and energy, such as our Sun, bend space and time around them, like weights sitting on a rubber sheet. Smaller objects are pulled into orbit around them by the "slope" they create.*

WOW!

After his death, Einstein's brain was taken away for research, without permission by a doctor at his hospital.

# Inside the atom

Until the 20th century, atoms were always thought to be the smallest things in the Universe – so small they could never be broken down into anything else. We now know that atoms are made up of even tinier particles – protons, neutrons, and electrons. These discoveries came about thanks to clever experiments carried out in the 1900s by scientists such as Ernest Rutherford and his colleagues.

GREAT DISCOVERIES

## ERNEST RUTHERFORD

Born in New Zealand in 1871, Ernest Rutherford studied in England, where he became a famous professor of physics and chemistry. In 1908, Rutherford successfully proved that radioactivity was the result of atoms breaking apart – something that had always been considered impossible. As well as making important discoveries, he also taught other physicists such as the Danish scientist Neils Bohr, who developed Rutherford's ideas. As a result, Rutherford is often known as the "father of nuclear physics".

## MODEL OF THE ATOM

Thanks to the British physicist J.J. Thomson, Rutherford and his students knew that the inside of an atom was made up of positive and negative particles (protons and electrons). They believed that these particles were all mixed up together, and designed an experiment to test their theory. To their surprise, the experiment instead showed that the positive particles inside an atom are all packed together in the centre, while the negative particles orbit around the outside.

▶ GOLD FOIL EXPERIMENT
*Positively charged radiation was fired at a thin sheet of gold. If the particles in the gold atoms were mixed together evenly, the radiation would pass through them easily without changing direction.*

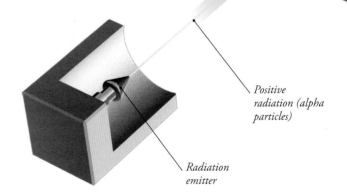

*Positive radiation (alpha particles)*

*Radiation emitter*

▲ LIFETIME IN RESEARCH *Ernest Rutherford worked and taught physics at Manchester University in the UK. He won the Nobel Prize for physics in 1908.*

282

## ANOTHER PARTICLE

Rutherford and his colleagues had proven the existence of protons and electrons, but they also knew that atoms were too heavy to be made up of just those particles. Rutherford came up with the idea of a third particle, called a neutron, with no electrical charge. It sits inside the nucleus of the atom, next to the protons. His idea remained a theory until 1932 when James Chadwick, a British physicist at the University of Cambridge in the UK, built a neutron detector to prove the existence of this third kind of particle.

Radiation passes to detector

Tube contains radioactive elements

▶ NEUTRON DETECTOR *Chadwick used two different radioactive elements, polonium and radium, to create a device that emitted neutrons.*

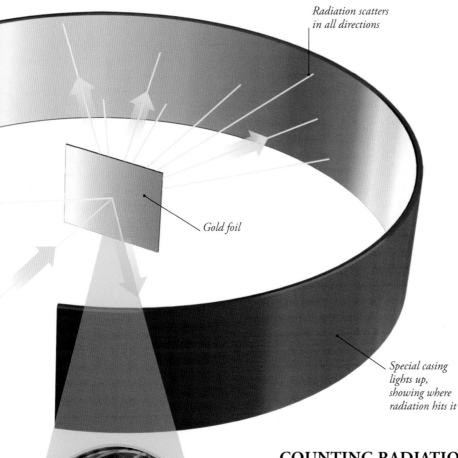

Radiation scatters in all directions

Gold foil

Special casing lights up, showing where radiation hits it

▲ NUCLEUS *The radiation was scattered in all directions. This showed that the positive parts of an atom are in the centre (nucleus), while the negative parts orbit around it, like planets around the Sun.*

### J.J. THOMSON

In 1897, Thomson became the first person to prove that atoms can be split into smaller particles. He passed electrons through a cathode ray tube, and discovered that magnets and static electricity caused the electrons to change direction. This showed that the electrons were tiny particles rather than waves of energy.

## COUNTING RADIATION

Another one of Rutherford's students was the German scientist Hans Geiger. In 1908, he devised a machine which could detect some types of radiation. Over the next 20 years, he and his student, Walther Müller, worked to build a machine that could detect all kinds of radioactivity. Their Geiger counter is still used around the world today.

▶ GEIGER COUNTER *This device is used across the world, for example to check for radioactive contamination in crops.*

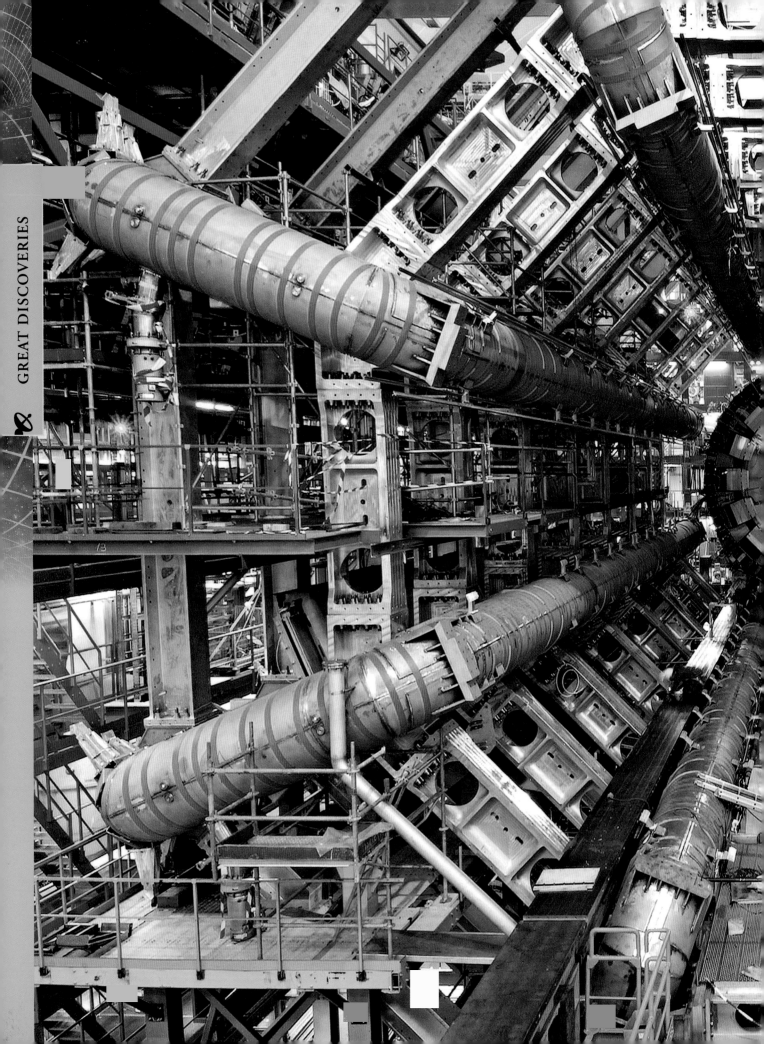

## ATOM SMASHER

The Large Hadron Collider uses powerful magnets to speed bits of atoms around a circular track 27 km (17 miles) long. The bits smash into each other at close to the speed of light, splitting into tiny particles. Scientists can study these particles to find out more about the Universe.

# The secret of life

How can something as complex as a human body grow from a single cell? The answer lies in deoxyribonucleic acid (DNA), a special kind of molecule hidden inside almost all the cells in your body. These long strings contain all the information needed to grow a human being. DNA was discovered in the 1940s, but nobody understood how it could carry genetic code. It was only in 1953 that scientists discovered its unique structure – a special spiral with information stored in its coils.

*DNA is coiled into X-shaped chromosomes inside the body's cells*

## CRACKING THE CODE

Francis Crick and James Watson were researchers at the University of Cambridge, England. Their research showed that DNA molecules are shaped in a two-part spiral (called a "double helix"), enabling them to carry a huge amount of information in a small space.

◀ DNA *These incredibly long molecules are hidden inside every cell of your body. Each one contains all the information needed for your body to grow.*

▲ SPIRAL MODEL *Watson (left) and Crick built a model of a DNA molecule in their laboratory in Cambridge, which they used to demonstrate their theory.*

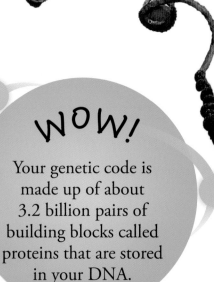

## WOW!

Your genetic code is made up of about 3.2 billion pairs of building blocks called proteins that are stored in your DNA.

## X-RAY PICTURES

Watson and Crick's discovery was made possible by the work of Rosalind Franklin. She and her colleague, Maurice Wilkins, used a new technique called X-ray crystallography to create the first clear images of DNA in 1952. These images gave Watson and Crick vital clues about the molecules' shape and structure.

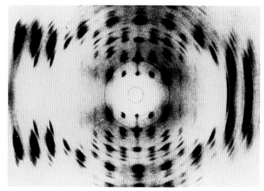

◀ MICROSCOPE VIEW
*Franklin produced images like this by reflecting X-rays off DNA molecules.*

▲ PAVING THE WAY *English scientist Rosalind Franklin produced the first clear images of DNA molecules, and identified some of the elements in them.*

## MAPPING THE GENES

In 1990, a project was launched to study every gene in a human being's genetic code. The idea was to work out what each gene was for, from defining the colour of a person's eyes to keeping them healthy. The project was completed in 2003, providing important information about how faulty genes can cause diseases.

*DNA stores genes in spirals.*

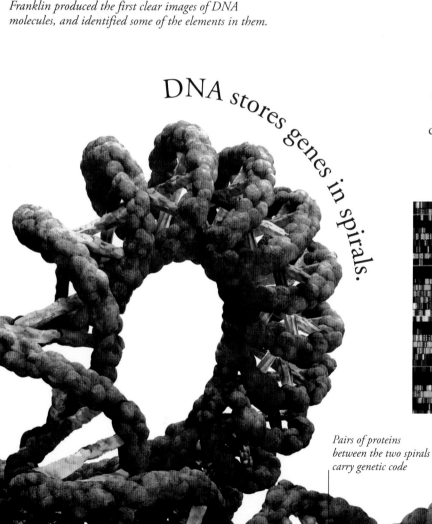

*Pairs of proteins between the two spirals carry genetic code*

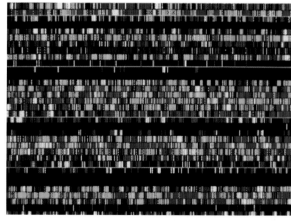

▲ HUMAN GENOME *These coloured squares map out the entire genetic code, or genome, of a person.*

# Great inventions

Inventions power progress – the never-ending quest to improve our world. Although some inventions occur by accident, most happen when science becomes technology. Science is the journey of discovery that reveals the wonder of the world, while technology means using science to solve human problems.

▼ **3000 BCE** BRONZE TOOLS
*A strong alloy (mixture) of copper and tin, bronze was easy to shape into tools, and was rustproof and very hard wearing.*

▼ **2 MILLION YEARS AGO** STONE TOOLS *The earliest stone tools were invented about 2 million years ago. These multi-purpose flint axes were held with one hand.*

► **c.5000 BCE** WRITING
*The Sumerians invented writing by pressing marks in clay tablets, so people could record crop harvests, taxes, and early business deals.*

► **2000 BCE** IRON TOOLS
*Iron, a common metal, is easy to hammer or "cast" into shape by melting and moulding.*

| 15000 BCE | 7500 BCE | 5000 BCE | 2500 BCE |
| --- | --- | --- | --- |

▼ **8000 BCE** FARMING *Early tribes moved around, hunting animals and gathering food. Modern life began when people settled, growing crops and farming animals instead.*

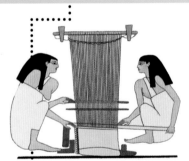

▲ **3000 BCE** COTTON FABRIC *The first clothes date from around 5,000 years ago. Cotton fabric was first made in the Indus Valley.*

▼ **5000 BCE** PLOUGH
*The first crops were planted with picks and digging sticks. Ploughs, pulled by animals, meant one person could work much more land.*

▲ **20,000 BCE** POTTERY *One of the oldest known handicrafts, pottery was invented when people discovered how to dig soft clay from the ground and "fire" it to make it hard.*

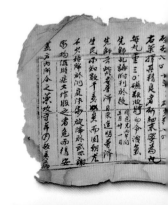

288

▲ **1430s PRINTING PRESS**
*After Johannes Gutenberg invented modern printing, books could be copied in their thousands. Before that, each volume had to be copied by hand.*

▲ **1700s FACTORY**
*The first factories used water and animal power to spin and weave textiles such as wool and cotton. Later ones were powered by steam.*

▲ **1608 TELESCOPE**
*Dutch spectacle-maker Hans Lippershey invented the telescope. Galileo Galilei improved it so he could watch the stars and planets.*

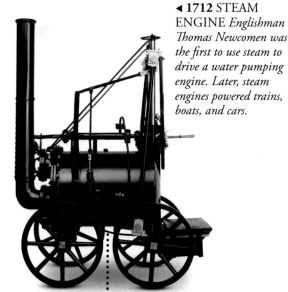

◄ **1712 STEAM ENGINE** *Englishman Thomas Newcomen was the first to use steam to drive a water pumping engine. Later, steam engines powered trains, boats, and cars.*

▲ **1876 TELEPHONE** *American scientist Alexander Graham Bell's telephone enabled people to talk over long distances for the first time.*

**100 CE** ⊢──────── **1500** ⊢──────────── **1800** ⊢────────

▲ **1280 SPECTACLES**
*English scientist Roger Bacon first came up with the idea of using a magnifying glass to help with reading.*

▼ **105 BCE PAPER** *Chinese inventor Ts'ai Lun made the first paper. Without its invention, printing would never have been possible.*

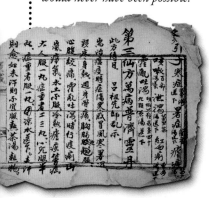

▼ **1590 MICROSCOPE**
*Microscopes were key to the development of biology. They helped scientists to discover cells, which power living things.*

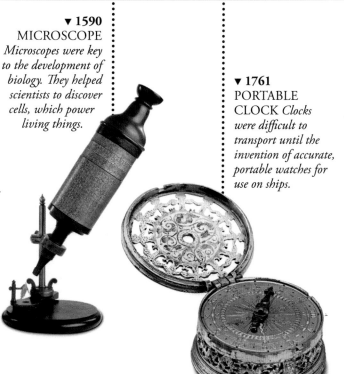

▼ **1761 PORTABLE CLOCK** *Clocks were difficult to transport until the invention of accurate, portable watches for use on ships.*

▼ **1878 LIGHT BULB**
*People had to make light by burning gas flames or by using candles until US inventor Thomas Edison perfected the electric light bulb.*

▼ **1880s** EARLIEST CINEMA *Frenchman Étienne-Jules Marey invented the movie camera. A decade later, Auguste and Louis Lumière showed the first cinema films.*

▲ **1903** AEROPLANE *Two American brothers, Wilbur and Orville Wright, invented the aeroplane when they added a petrol-powered engine to a giant glider.*

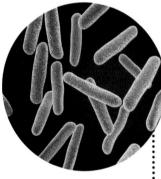

▲ **1931** ELECTRON MICROSCOPE *Germans Max Knoll and Ernst Ruska built the first microscope that could see ultra-small things.*

▲ **1928** ANTIBIOTICS *Scottish biologist Alexander Fleming helped to save millions of lives when he discovered penicillin, the first antibiotic drug.*

⊢ **1880** ⊢        **1900** ⊢

▶ **1890** RADIO BROADCASTS *Although Italian Guglielmo Marconi did not invent radio, he demonstrated how it could send messages around the world.*

▶ **1926** TELEVISION *Scottish inventor John Logie Baird used radio waves to send pictures, giving the first demonstration of television.*

▼ **1885** CAR *German engineer Karl Benz built the first primitive car when he attached a petrol-powered engine to a three-wheel carriage.*

▼ **1937** JET ENGINE *Englishman Frank Whittle developed a faster engine that could push aeroplanes forward using a hot jet of burning gas.*

◀ **1973** MOBILE PHONES *Radio telephones date back to the 1920s, but the first mobile phone call was not made until 1973, by Martin Cooper of Motorola Corporation.*

▼ **1996** CLONED SHEEP *Dolly the sheep became world famous as the first animal "made" by cloning – copying genetic material in a laboratory.*

▼ **1974** INTERNET *Americans Vinton Cerf and Bob Kahn invented a way of sending packets of information between different computers.*

▲ **1969** MAN ON THE MOON *US astronauts Neil Armstrong and Edwin ("Buzz") Aldrin became the first people to walk on the Moon.*

1950 ——— 1970 ——————————————————— 2005

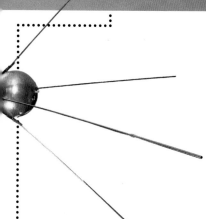

▼ **1977** PERSONAL COMPUTERS *Early computers filled entire rooms. Smaller versions, such as the Apple II and Commodore PET (below), brought portable computer power to people's homes.*

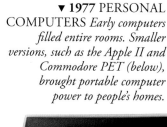

▲ **1982** ARTIFICAL HEART *Dr Robert Jarvik developed the first mechanical heart in 1982. Made from plastic and aluminium, it was powered by air pressure.*

▲ **1957** SPACE SATELLITE *The Soviet Union started the space race by launching Sputnik 1, the first artificial satellite.*

▲ **2004** SMART PHONE *Computers and mobile phones merged together when smart phones were invented. Their touch screens work as both displays and keyboards.*

291

# Glossary

**Acceleration** The change in the velocity (speed and direction) of a moving object.

**Air pressure** The force made by air molecules pushing against a surface or container.

**Alloy** A material made by mixing a metal with small amounts of other metals or non-metals. Alloys are stronger, harder, or more useful than the metal they are based on.

**Android** A type of robot that looks and behaves like a human being.

**Astronomy** The study of objects in space, including stars, planets, and galaxies.

**Atom** The smallest possible part of a chemical element. Atoms are made of protons, neutrons, and electrons.

**Atomic mass** The total number of protons and neutrons that an atom contains.

**Atomic number** The number of protons an atom contains.

**Barometer** A scientific instrument that measures air pressure. Some barometers also show simple weather forecasts.

**Battery** A portable electricity supply that stores electric charge using chemicals.

**Big Bang** A powerful explosion from which our Universe was created about 14 billion years ago.

**Boiling point** The temperature at which a liquid starts to change into a gas.

**Buoyancy** The upward force on an object in a liquid, caused by the water pressure underneath it. Buoyancy pushes against the object's weight and tends to make it float.

**Camouflage** A surface pattern, body shape, or colour that living things use to hide themselves from predators.

**Carbohydrate** A chemical compound, found in starchy foods such as rice, potatoes, and bread, which gives us energy. Carbohydrates are made from carbon, hydrogen, and oxygen.

**Carnivore** An animal that eats the meat of other animals. Plants that eat insects are also sometimes called carnivores.

**Cell** The basic unit from which all living organisms are made.

**Centigrade** A temperature scale based on the melting point of ice (0°C) and the boiling point of water (100°C), with 100 equal divisions, called degrees, in between them.

**Chemical** A substance made from elements or compounds, which are made from atoms and molecules.

**Chlorophyll** A green chemical found in plants. With the help of sunlight, chlorophyll helps plants turn water and carbon dioxide into food by photosynthesis.

**Chloroplast** A part of a plant cell that contains chlorophyll.

**Chromosome** A structure in the nucleus of a cell, made from coiled DNA strands, that carries genetic information.

**Circuit** A path that electricity flows around. All electrical and electronic things have circuits inside them.

**Climate change** Long-term changes in Earth's weather patterns. These changes can result from orbital cycles, global environmental variations, and in recent times, human activity.

**Clone** An organism that has exactly the same genetic material as its parent.

**Cold-blooded** An animal whose body temperature changes to match the temperature of their surroundings.

**Colony** A group of organisms of the same species that live together to support and protect one another.

**Combustion** A chemical reaction in which a fuel, such as wood or coal, burns with oxygen from the air to release heat energy.

**Compound** A chemical made by combining the atoms of two or more different elements.

▶ HUDDLE UP *Animals living in extremely cold habitats stay together to keep warm. Huddling allows conduction of heat, in this case from the mother to the baby penguin.*

**Compound eyes** The eyes in creatures such as flies, made from thousands of separate light-detecting units called ommatidia.

**Concave lens** A lens in which one or both outer edges curve ("cave") inwards. Concave lenses tend to make light spread out, or diverge, so they are also called diverging lenses.

**Condensation** The process by which a gas or vapour turns into a liquid when the temperature falls or the pressure rises.

**Conduction** The flow of electricity or heat through a material.

**Convection** The process by which heat spreads through a liquid or gas as warmer material floats to the surface.

**Convex lens** A lens in which one or both the outer edges curve outwards. Convex lenses tend to make light focus on a point, or converge, so they are also called converging lenses.

**Diode** An electronic component that allows an electric current to flow through a circuit in only one direction.

**DNA** Deoxyribonucleic acid. The chemical inside chromosomes that lets parents pass genetic information on to their offspring.

**Ecosystem** A community of living plants and animals, and non-living things such as air or water, that occupy the same habitat and interact with one another.

**Elasticity** A property of a material that allows it to stretch or bend when you push or pull it and then return to its original shape and length.

**Electricity** A type of energy caused by electrons inside atoms. Static electricity is made by electrons building up in one place, while current electricity happens when electrons move around.

▲ HOW BIG? *Our Universe is incredibly huge. It includes planets, stars, galaxies, and dust clouds. The Earth is a tiny dot in this vast space.*

**Electrode** An electrical contact, made from a conductor, that connects the main part of a circuit to something outside it, such as the chemicals in a battery.

**Electromagnet** A coil of wire that makes temporary magnetism when electricity flows through it.

**Electron** A tiny particle with a negative electric charge that occupies the empty outer space around an atom's nucleus. Moving electrons can carry heat or electricity and make magnetism.

**Electronics** A way of using small, very precise amounts of electricity to carry information, such as TV pictures or computer data, or to control electric appliances.

**Element** A basic building block of matter made from identical atoms.

**Energy** A property of an object that allows it to do something now or in the future. Different types of energy include kinetic energy (movement energy) and potential energy (stored energy).

**Engine** A device that burns fuel and oxygen to release stored heat energy that can power a machine.

**Enzyme** A catalyst that living things use to make chemical reactions happen more quickly inside them.

**Evaporation** The process by which a liquid turns into a vapour when the temperature rises or the pressure falls.

**Exoskeleton** A hard skeleton on the outside of an insect or crustacean that protects the body inside.

**Extinct** A species that has completely died out, so no members of it are still alive anywhere on Earth.

**Fahrenheit** A temperature scale with the melting point of ice at 32°F and the boiling point of water at 212°F, originally proposed by German physicist Daniel Fahrenheit.

**Filament** A thin coil of wire that glows as electricity flows through it. A filament makes the light in an electric lamp.

**Force** A pushing or pulling action that changes an object's speed, direction of movement, or shape.

**Fossil** The remains of a plant or animal preserved inside a rock.

**Fossil fuel** A fuel made mostly from carbon and hydrogen that can be burned with oxygen to release heat energy. Coal, oil, and gas are fossil fuels.

**Freezing point** The temperature at which a liquid turns into a solid.

**Frequency** A measurement of how often a wave of energy moves up and down. Waves with a higher frequency move up and down faster.

**Friction** The rubbing force between two things that move past one another. Friction slows things down and generates heat.

**Galaxy** A large group of stars, dust, and gas held together by the force of gravity. We live in a galaxy called the Milky Way.

**Gear** One of a pair of wheels of different sizes, with teeth cut into their edges, that turn together to increase the speed or force of a machine.

**Genome** The complete collection of genetic information inside a living thing.

**Global warming** A very gradual rise in the average temperature of Earth's atmosphere and oceans, thought to be caused by people burning fossil fuels.

**Gravity** The pulling force between every mass in the Universe and every other mass.

**Habitat** The place where a plant or an animal normally lives.

**Herbivore** An animal that eats plants.

**Ice Age** A period of history when Earth's atmosphere and oceans were much cooler and more of the planet was covered by glaciers and ice sheets.

**Infrared** A type of electromagnetic radiation that carries energy in invisible beams from hot objects.

**Laser** Light Amplification by Stimulated Emission of Radiation. A very powerful beam of light made by exciting atoms inside a tube.

**Lens** A curved, transparent piece of plastic or glass that can bend light rays to make something look bigger, smaller, closer, or further away.

**Lever** A rod balanced on a pivot that can increase the size of a pushing, pulling, or turning force. Crowbars, bottle openers, and seesaws all work as levers.

**Light-emitting diode (LED)** A small electronic component that converts electricity into light without making heat.

**Light year** The distance light travels in a year. One light year is about 9.5 trillion km (6 trillion miles).

◄ EARLY CALENDARS
*Ancient civilizations often determined seasons, months, and years based on the movements of the Sun, Moon, and stars. This Aztec calendar was based on the Sun's cycles.*

**Magnetic field** The invisible pattern of force that stretches out around a magnet.

**Mass** The amount of matter that an object contains.

**Matter** The material which everything around us is made from. Matter includes solids, liquids, and gases and both living and non-living things.

**Melting point** The temperature at which a solid turns into a liquid.

**Metamorphosis** The process by which a living thing changes into a different form, such as when a caterpillar changes into a butterfly.

**Microchip** A miniature circuit made from thousands, millions, or even billions of separate electronic components. The components are so small that the chip is usually no bigger than a fingernail.

**Microscope** A scientific instrument that uses lenses and mirrors to make small objects appear much larger.

**Molecule** A substance made from two or more atoms bonded (joined tightly) together. The atoms in a molecule might be the same or different.

**Motor** A machine that uses electricity and magnetism to produce spinning movement or sometimes movement in a straight line.

**Nanotechnology** A way of building tiny materials by joining together individual atoms or molecules.

**Natural selection** The process by which a species evolves through time by passing on useful improvements to the next generation.

**Neutron** A particle inside the nucleus of an atom that has a neutral electric charge.

▲ LIGHT TRAVEL *A fibre-optic cable is created from thin strands of optical fibre, which are made of plastic or glass. These cables are used for quick transfer of information such as sound, pictures, or computer code, in the form of light.*

**Nuclear fission** A process in which large atoms break into smaller ones, giving off large amounts of energy.

**Nuclear fusion** A process in which small atoms, such as hydrogen, join together to make larger ones, such as helium, and give off large amounts of energy. The Sun is powered by nuclear fusion.

**Nucleus** The central part of an atom made from protons and neutrons.

**Omnivore** An animal that feeds on both plants and animals.

**Organic compound** A compound made from carbon and other elements.

**Parasite** An organism that lives off another (usually much larger) organism.

**pH** A measurement that tells us whether an acid or base is strong or weak.

**Photosynthesis** The process by which growing plants use sunlight, water, and carbon dioxide to make food.

**Polymer** An organic molecule made from many identical units (monomers) joined together. Plastics are examples of polymers.

**Predator** An animal that hunts other animals for food.

**Pressure** The force pushing on a surface. The bigger the force or the smaller the area it acts on, the higher the pressure.

**Prey** An animal hunted or eaten by a predator.

**Proton** A particle inside the nucleus of an atom that has a positive electric charge.

**Radiation** 1: A burst of particles or energy released by an unstable atom. 2: A process by which heat transfers through air or empty space. 3: An electromagnetic wave.

**Reflection** The way light, sound, or other types of energy bounce back from a surface.

**Refraction** The way light bends and changes direction when it passes from one material into another.

**Renewable energy** A type of energy that will not run out, generated from sources such as wind, waves, and sunlight.

**Resistor** An electronic component that reduces the electric current flowing through a circuit.

**Satellite** An object in space that travels around another in a path called an orbit. Artificial satellites placed into Earth's orbit by humans take photographs, transmit data, and help us navigate.

**Species** A group of similar organisms that can breed with one another to produce offspring.

**Transistor** An electronic component that can make electric currents bigger or switch them on and off. Transistors are the basic components in computer memory and microchips.

**Transmitter** A device that turns electricity into radio waves and broadcasts them through air or space.

**Ultrasound** A type of high-frequency sound that humans cannot hear. Ultrasound is used for medical scanning and testing materials.

**Vacuum** An empty space that contains no air or other matter.

**Warm-blooded** An animal whose body temperature stays almost constant even if its surroundings grow hotter or colder. Humans and birds are warm-blooded.

**Wavelength** The distance between the peaks of a wave.

▶ EXPLORING SPACE *The first space shuttle was launched in 1981. This is the US space shuttle Endeavour at NASA's Kennedy Space Center in Florida, USA, launched in 2009.*

# Index

REFERENCE

# Acknowledgements

DK would like to thank David Burnie for consultancy advice, Joanna Shock for proofreading, Elizabeth Wise for the index, Olivia Stanford and Lizzie Davey for editorial assistance, Surya Sarangi and Sakshi Saluja for picture research assistance, and Mrinal Duggal for illustrations.

The publisher would like to thank the following for their kind permission to reproduce their photographs:

(Key: a-above; b-below/bottom; c-centre; f-far; l-left; r-right; t-top)

2–3 **Dreamstime.com:** Brett Critchley. 4 **Dorling Kindersley:** NASA (cr). **Dreamstime.com:** Corepics Vof (crb). 5 **123RF.com:** Roman Gorielov (cb/Paints). **Corbis:** Mediscan (crb/Bacteria). **Dorling Kindersley:** Malcolm Coulson (cr); The Royal British Columbia Museum, Victoria, Canada (cra/Mammoth); The Senckenberg Nature Museum, Frankfurt (br). **Dreamstime.com:** Anikasalsera (c); Felix

Miznieznikov (ca/Amusement Park Ride); Ben Heys (cb/Water Drop); Štěpán Kápl (cb/Solar Power); Anthony Bolan (tr/Pixel Pattern). **Getty Images:** Jacom Stephens / E+ (tc). **NASA:** JPL-Caltech (ca/Nebula). **Science Photo Library:** Alfred Benjamin (bc); Dr. Charles Mazel / Visuals Unlimited, inc (cra/Mouse); Pasieka (crb/Micrograph). 6–7 **Corbis:** Dan McCoy - Rainbow / Science Faction. 6–57 **Science Photo Library:** Frans Lanting, Mint Images (sidebar). 7 **Corbis:** Martin Rietze / Westend61 (ca). **Dorling Kindersley:** The Natural History Museum, London (cla/nugget). **Dreamstime.com:** Diego Cervo (br). 8 **Dreamstime.com:** Snapgalleria (c). **NASA:** ESA / ASU / J. Hester (tr). 8–9 **Corbis:** Pat O'Hara. 9 **Dorling Kindersley:** The Natural History Museum, London (tl); The Science Museum, London (br). **Getty Images:** Photodisc / John William Banagan (cr). 11 **Science Photo Library:** Prof. Erwin Mueller (b). 12–13 **Science Photo Library:** Pasieka. 13 **Dreamstime.com:** Dbjohnston (tc); Sergei Primakov (c).

14 **Corbis:** Long Wei / epa (bl). **Dreamstime.com:** Jianghongyan (bc). **Pearson Asset Library:** Trevor Clifford / Pearson Education Ltd (br). 15 **Alamy Images:** Jim West (tr). 16–17 **Science Photo Library:** Javier Trueba / MSF. 18 **Getty Images:** Roberto la Forgia / Flickr (bl). 18–19 **Science Photo Library:** Paul Rapson. 19 **Dreamstime.com:** Marcel Clemens (cra). 20 **Alamy Images:** ImageDJ (bl). 20–21 **Dreamstime.com:** Hel080808 (t). 21 **Corbis:** Olaf Döring / imagebroker (clb). **Fotolia:** reflektastudios (cla). 24 **Dreamstime.com:** Artem Gorohov (bl); Eimantas Buzas (br). **Pearson Asset Library:** Coleman Yuen / Pearson Education Ltd (clb). 25 **Corbis:** Theodore Gray, / Visuals Unlimited (br). **Science Photo Library:** (bc). 27 **Dreamstime.com:** Jörg Beuge (c). **Pearson Asset Library:** Cheuk-king Lo / Pearson Education Ltd (bc); Joey Chan / Pearson Education Ltd (bl); Gareth Boden / Pearson Education Ltd (bc/jar). **Science Photo Library:** Charles D. Winters (clb). 28

**Dreamstime.com:** Diego Cervo (l). 29 **Corbis:** Christopher Berkey / epa (br). **Dreamstime.com:** Denise Peillon (cl). **Science Photo Library:** Charles D. Winters (tl). 30 **Pearson Asset Library:** Studio 8 / Pearson Education Ltd (cb). 31 **Dreamstime.com:** Valentyn Volkov (c). **Science Photo Library:** Ted Kinsman (tr). 32 **Dreamstime.com:** Christina Richards (cl). 33 **Dreamstime.com:** Lkordela (cra). **Getty Images:** fStop Images – Caspar Benson / Brand X Pictures (br). 34–35 **Science Photo Library:** Ria Novosti. 36 **Dreamstime.com:** Rudy Umans (clb). 37 **Dorling Kindersley:** The University Museum of Archaeology and Anthropology, Cambridge (bl/bowl); The Natural History Museum, London (ftr); The Museum of London (bl/cup, bl/spear-tip). **Dreamstime.com:** Daniel Nagy (crb); Somyot Pattana (tr). **Science Photo Library:** Martyn F. Chillmaid (c). 38–39 **Dorling Kindersley:** The March Field Air Museum, California (bl). 39 **Alamy Images:** Dave Marsden (cr); Thomas Jackson (tl). **Dreamstime.com:**

REFERENCE

**REFERENCE**